State-of-the-Art Deep Learning Models in TensorFlow

Modern Machine Learning in the Google Colab Ecosystem

David Paper

Apress®

State-of-the-Art Deep Learning Models in TensorFlow: Modern Machine Learning in the Google Colab Ecosystem

David Paper
Logan, UT, USA

ISBN-13 (pbk): 978-1-4842-7340-1 ISBN-13 (electronic): 978-1-4842-7341-8
https://doi.org/10.1007/978-1-4842-7341-8

Managing Director, Apress Media LLC: Welmoed Spahr
Acquisitions Editor: Jonathan Gennick
Development Editor: Laura Berendson
Coordinating Editor: Jill Balzano
Technical Reviewer: Brendan Tierney

Cover designed by eStudioCalamar

Cover image designed by Pexels

Distributed to the book trade worldwide by Springer Science+Business Media LLC, 1 New York Plaza, Suite 4600, New York, NY 10004. Phone 1-800-SPRINGER, fax (201) 348-4505, e-mail orders-ny@springer-sbm.com, or visit www.springeronline.com. Apress Media, LLC is a California LLC and the sole member (owner) is Springer Science + Business Media Finance Inc (SSBM Finance Inc). SSBM Finance Inc is a **Delaware** corporation.

For information on translations, please e-mail booktranslations@springernature.com; for reprint, paperback, or audio rights, please e-mail bookpermissions@springernature.com, or visit http://www.apress.com/rights-permissions.

Apress titles may be purchased in bulk for academic, corporate, or promotional use. eBook versions and licenses are also available for most titles. For more information, reference our Print and eBook Bulk Sales web page at http://www.apress.com/bulk-sales.

Any source code or other supplementary material referenced by the author in this book is available to readers on GitHub via the book's product page, located at www.apress.com/978-1-4842-7340-1. For more detailed information, please visit http://www.apress.com/source-code.

Printed on acid-free paper

I dedicate my book to my Mom and younger brother Bruce. Both have been instrumental in my development as an author and a person.

Table of Contents

About the Author xxi

Introduction xxiii

Chapter 1: Build TensorFlow Input Pipelines 1

What Are Input Pipelines? 1

Why Build Input Pipelines? 2

Manual Workflow 2

Automated Workflow 3

Basic Input Pipeline Mechanics 3

High-Performance Pipelines 4

Google Developers Codelabs 4

Create a New Notebook in Colab 4

Import the TensorFlow Library 5

GPU Hardware Accelerator 5

Create a TensorFlow Dataset 6

Consume the Dataset 7

Dataset Structure 8

Create a Dataset from Memory 8

Load and Inspect Data 8

Scale and Create the tf.data.Dataset 9

Verify Scaling 11

Check Tensor Shape 11

Inspect Tensors 11

Preserve the Shape 11
Visualize 12
Define Class Labels 12
Convert a Numerical Label to a Class Label 12
Create a Plot of Examples from the Dataset 13
Build the Consumable Input Pipeline 13
Build the Model 15
Compile and Train the Model 17
Load a TensorFlow Dataset as NumPy 18
Inspect Shapes and Pixel Intensity 19
Scale 19
Prepare Data for TensorFlow Consumption 20
Build the Consumable Input Pipeline 20
Build the Model 20
Compile and Train the Model 21
Create a Dataset from Files 21
Download and Inspect the Dataset 22
Parse Data with the tf.keras.preprocessing Utility 24
Get Flowers from Google Cloud Storage 29
Read and Process Flowers as TFRecord Files 31
Summary 36

Chapter 2: Increase the Diversity of Your Dataset with Data Augmentation 37

Data Augmentation 37
Import the TensorFlow Library 38
GPU Hardware Accelerator 38
Augment with a Keras API 39
Get Data 39
Split Data 40
Inspect the Training Set 41
Get Number of Classes 41
Create a Scaling Function 42

Build the Input Pipeline 42
Data Augmentation with Preprocessing Layers 42
Display an Augmented Image 43
Create the Model 44
Compile and Train the Model 45
Visualize Performance 46
Apply Augmentation on Images 46
Set an Index Variable 47
Set an Image 47
Show an Example 47
Create Functions to Show Images 48
Crop an Image 49
Randomly Flip Image Left to Right 50
Randomly Flip Image Up to Down 50
Flip Image Up to Down 50
Rotate Image 90 Degrees 50
Adjust Gamma 50
Adjust Contrast 51
Adjust Brightness 52
Adjust Saturation 52
Hue 52
Apply Transformations Directly on Images 53
Data Augmentation with ImageGenerator 56
Process Flowers Data 56
Create Datasets 57
Create the Model 58
Compile and Train the Model 59
Augment Training Data 59
Recompile and Train the Model 60
Inspect the Data 61
Visualize 62
Summary 64

Chapter 3: TensorFlow Datasets 65
An Introduction to TensorFlow Datasets 65
Import the TensorFlow Library 66
GPU Hardware Accelerator 66
Available Datasets 66
Load a Dataset 67
TFDS Metadata 68
Iterate Over a Dataset 68
As a Dictionary 68
As Tuples 69
As NumPy Arrays 69
Visualization 71
tfds.as_dataframe 71
Take Examples 71
tfds.show_examples 72
Load Fashion-MNIST 72
Metadata 73
Visualize 74
Slicing API 75
Slicing Instructions 76
Performance Tips 80
Auto-caching 80
Benchmark Datasets 80
Reloading a TFDS Object 81
Load Fashion-MNIST as a Single Tensor 81
Ready Data for TensorFlow Consumption 82
Build the Input Pipeline 82
Build the Model 82
Compile and Train the Model 83
Load Beans as a tf.data.Dataset 84
Metadata 85

Visualize 85
Check Shapes 87
Reformat Images 87
Configure Dataset for Performance 87
Build the Model 88
Compile and Train the Model 90
Predict 90
Summary 91
Chapter 4: Deep Learning with TensorFlow Datasets 93
An Experiment with cats_vs_dogs 93
Import the TensorFlow Library 94
GPU Hardware Accelerator 94
Begin the Experiment 94
Load the TFDS Object 95
Metadata 95
Split the Data 96
Visualize 97
Inspect Examples 98
Reformat Images 99
Build the Input Pipeline 99
Visualize and Inspect Examples from a Batch 100
Build the Model 101
Compile and Train the Model 103
Evaluate the Model for Generalizability 104
Visualize Performance 104
Augmentation with Preprocessing Layers 105
Build the Model 106
Compile and Train the Model 107
Evaluate the Model for Generalizability 107
Visualize Performance 108

Apply Data Augmentation on Images 108
Build the Input Pipeline 109
Display an Augmented Image 109
Build the Model 110
Compile and Train the Model 110
Evaluate the Model for Generalizability 111
Visualize Performance 111
Predictions 111
Visualize Predictions 113
An Experiment with rock_paper_scissors 115
Configure TensorBoard 115
Load Data 116
Inspect the Data 116
Preprocess the Data 117
Visualize Processed Data 118
Augment Training Data 118
Augment Train Data 121
Build the Input Pipeline 121
Create the Model 122
Compile and Train 123
Visualize Performance 124
Close TensorBoard Server 125
Chapter 5: Introduction to Tensor Processing Units 127
TPU Hardware Accelerator 127
Advantages of Cloud TPU 128
When to Use Cloud TPU 128
Import the TensorFlow Library 129
Enable TPU Runtime 129
TPU Detection 130
Configure the TPU for This Notebook 130
Create a Distribution Strategy 131

Manual Device Placement.......... 131
Run a Computation in All TPU Cores.......... 132
Eager Execution.......... 132
Experiments.......... 133
Digits Experiment.......... 133
    Preprocess the Data.......... 134
    Build the Input Pipeline.......... 135
    Model Data Within TPU Scope.......... 136
MNIST Experiment.......... 137
    Build the Input Pipeline.......... 138
    Model Data Within TPU Scope.......... 138
Fashion-MNIST Experiment.......... 140
    Transform Datasets into Image and Label Sets.......... 141
    Model Data Within TPU Scope.......... 141
    Save the Trained Model.......... 143
    Make Inferences.......... 143
Flowers Experiment.......... 145
    Read Flowers Data as TFRecord Files.......... 146
    Set Parameters for Training.......... 147
    Create Functions to Load and Process TFRecord Files.......... 148
    Create Train and Test Sets.......... 150
    Model Data.......... 150
    Make Inferences.......... 151
Chapter 6: Simple Transfer Learning with TensorFlow Hub.......... 153
Pre-trained Models for Transfer Learning.......... 154
Import the TensorFlow Library.......... 154
GPU Hardware Accelerator.......... 154
TensorFlow Hub.......... 155
MobileNet-v2.......... 155
Flowers MobileNet-v2 Experiment.......... 156

Load Flowers as a TFDS Object ... 156
Explore Metadata ... 156
Display Images and Shapes ... 157
Build the Input Pipeline ... 158
Create a Feature Vector ... 158
Freeze the Pre-trained Model ... 159
Create a Classification Head ... 159
Compile and Train the Model ... 160
Visualize Performance ... 161
Make Predictions from Test Data ... 162
Plot Predictions ... 164
Flowers Inception-v3 Experiment ... 164
Build the Input Pipeline ... 165
Build the Model ... 166
Compile and Train ... 166
Visualize Performance ... 167
Predictions ... 168
Plot Model Predictions ... 168
Summary ... 169
Chapter 7: Advanced Transfer Learning ... 171
Transfer Learning ... 171
Import the TensorFlow Library ... 172
GPU Hardware Accelerator ... 173
Beans Experiment ... 173
Load Beans ... 174
Explore the Data ... 174
Visualize ... 175
Reformat Images ... 175
Build the Input Pipeline ... 175
Model with Xception ... 176
Model Beans with Inception ... 180

Stanford Dogs Experiment 183
Model Stanford Dogs with MobileNet 183
Load Data 184
Metadata 184
Visualize Examples 184
Check Image Shape 186
Build the Input Pipeline 186
Create the Model 187
Compile and Train 188
Model Trained Data with Unfrozen Layers 190
Generalize 190
Flowers Experiment 190
Read Flowers as TFRecords 191
Create Data Splits 191
Create Functions to Load and Process TFRecord Files 192
Create Training and Test Sets 193
Create the Model 194
Compile and Train 195
Visualize 196
Generalize 196
Rock-Paper-Scissors Experiment 196
Load the Data 196
Visualize 197
Build the Input Pipeline 197
Create the Model 198
Compile and Train 198
Visualize 199
Generalize 199
Tips and Concepts 199

Chapter 8: Stacked Autoencoders 201
Import the TensorFlow Library 202
GPU Hardware Accelerator 202
Basic Stacked Encoder Experiment 203
Load Data 203
Scale Data 203
Build the Stacked Encoder Model 203
Compile and Train 205
Visualize Performance 205
Visualize the Reconstructions 206
Breakdown 208
Dimensionality Reduction 208
Tying Weights Experiment 210
Define a Custom Layer 210
Build the Tied Weights Model 211
Compile and Train 212
Visualize Training Performance 213
Visualize Reconstructions 213
Denoising Experiment 213
Build the Denoising Model 213
Compile and Train 214
Visualize Training Performance 215
Visualize Reconstructions 215
Dropout Experiment 215
Build the Dropout Model 216
Compile and Train 217
Visualize Training Performance 217
Visualize Reconstructions 217
Summary 217

Chapter 9: Convolutional and Variational Autoencoders 219
Import the TensorFlow Library 220
GPU Hardware Accelerator 220
Convolutional Encoder Experiment 220
Load Data 221
Inspect Data 221
Display Examples 221
Get Training Data 222
Inspect Shapes 223
Preprocess Image Data 223
Create a Convolutional Autoencoder 223
Compile and Train 225
Visualize Training Performance 225
Visualize Reconstructions 226
Variational Autoencoder Experiment 227
Load Data 228
Inspect Data 228
Scale 228
Create a Custom Layer to Sample Codings 229
Create the VAE Model 229
Compile and Train 231
Visualize Reconstructions 232
Generate New Images 232
TFP Experiment 234
Create a TFP VAE Model 236
Compile and Train 239
Efficacy Test 240
Summary 241

Chapter 10: Generative Adversarial Networks 243
Import the TensorFlow Library 244
GPU Hardware Accelerator 244
GAN Experiment 245
Load Data 245
Scale 245
Build the GAN 246
Compile the Discriminator Model 248
Build the Input Pipeline 248
Create a Custom Loop for Training 249
Train the GAN 251
DCGAN Experiment with Small Images 252
Create the Generator 252
Create the Discriminator 254
Create the DCGAN 254
Compile the Discriminator Model 254
Reshape 255
Build the Input Pipeline 255
Train 255
Generate Images with the Trained Generator 256
DCGAN Experiment with Large Images 256
Inspect Metadata 256
Load Data for Training 257
Massage the Data 257
Build the Input Pipeline 258
Build the Model 259
Compile the Discriminator Model 261
Rescale 261
Train Model and Generate Images 262
Summary 263

Chapter 11: Progressive Growing Generative Adversarial Networks 265
Latent Space Learning 265
Import the TensorFlow Library 266
GPU Hardware Accelerator 267
Create Environment for Experiments 267
Install Packages for Creating Animations 267
Install Libraries 268
Create Functions for Image Display 269
Create Latent Space Dimensions 269
Set Verbosity for Error Logging 270
Image Generation Experiment 270
Create a Function to Interpolate a Hypersphere 271
Load the Pre-trained Model 272
Generate and Display an Image 273
Create a Function to Generate Multiple Images 273
Create an Animation 274
Display Interpolated Image Vectors 274
Display Multiple Images from a Single Vector 276
Create a Target Latent Vector from an Uploaded Image 277
Custom Loop Learning Experiment 278
Create the Feature Vector 278
Display an Image from the Feature Vector 279
Create the Target Image 279
Create a Function to Find the Closest Latent Vector 280
Create the Loss Function 281
Train the Model 282
Training Loss 282
Animate 283
Compare Learned Images to the Target 283
Try a Different Loss Function Algorithm 284

Create a Target from an Uploaded Image 285
Create a Target from a Google Drive Image 287
Create a Target from a Wikimedia Commons Image 290
Latent Vectors and Image Arrays 292
Generate a New Image from a Latent Vector 292
Generate a New Image from an Image Vector 293
Summary 294
Chapter 12: Fast Style Transfer 295
Why Is Style Transfer Important? 296
Arbitrary Neural Artistic Stylization 296
Import the TensorFlow Library 298
GPU Hardware Accelerator 298
ANAS Experiment 298
Import Requisite Libraries 299
Get Images from Google Drive 299
Preprocess Images 301
Display Processed Images 301
Prepare Image Batches 302
Load the Model 302
Feed the Model 303
Explore the Pastiche 303
Visualize 305
Image Stylization with Multiple Images 306
TensorFlow Lite Experiment 311
Architecture for a Pre-trained TensorFlow Lite Model 311
Crop Images 312
Stylize Images 313
Create the Pastiche 316
Style Blending 316
Save the Pastiche 318
Summary 319

Chapter 13: Object Detection 321
Object Detection in a Natural Scene 321
Detection vs. Classification 322
Bounding Boxes 323
Basic Structure 324
Import the TensorFlow Library 325
GPU Hardware Accelerator 325
Object Detection Experiment 326
Import Requisite Libraries 327
Create Functions for the Experiment 328
Load a Pre-trained Object Detection Model 331
Load an Image from Google Drive 332
Prepare the Image 333
Run Object Detection on the Image 333
Detect Images from Complex Scenes 335
Summary 339
Chapter 14: An Introduction to Reinforcement Learning 341
Challenges of Reinforcement Learning 342
Import the TensorFlow Library 344
GPU Hardware Accelerator 344
Reinforcement Learning Experiment 344
Install and Configure OpenAI Gym on Colab 345
Import Libraries 346
Create an Environment 346
Display the Rendering from the Environment 347
Display Actions 348
Simple Neural Network Reward Policy 351

Model Predictions 352
Animate 354
Implement a Basic Reward Policy 355
Reinforce Policy Gradient Algorithm 357
Summary 364

Index 365

About the Author

Dr. David Paper is a retired academic from the Utah State University (USU) Data Analytics and Management Information Systems Department in the Huntsman School of Business. He has over 30 years of higher education teaching experience. At USU, he taught for 27 years in the classroom and through distance education over satellite. He taught a variety of classes at the undergraduate, graduate, and doctorate levels, but he specializes in applied technology education.

Dr. David Paper has competency in several programming languages, but his focus is currently on deep learning with Python in the TensorFlow-Colab ecosystem. He has published extensively on machine learning (ML) including such books as *Data Science Fundamentals for Python and MongoDB* (2018, Apress), *Hands-on Scikit-Learn for Machine Learning Applications: Data Science Fundamentals with Python* (2019, Apress), and *TensorFlow 2.x in the Colaboratory Cloud: An Introduction to Deep Learning on Google's Cloud Service* (2021, Apress). He has also published more than 100 academic articles.

Besides growing up in family businesses, Dr. Paper has worked for Texas Instruments, DLS Inc., and the Phoenix Small Business Administration. He has performed information systems (IS) consulting work for IBM, AT&T, Octel, the Utah Department of Transportation, and the Space Dynamics Laboratory. He has worked on research projects with several corporations, including Caterpillar, Fannie Mae, Comdisco, IBM, Raychem, Ralston Purina, and Monsanto. He maintains contacts in corporations such as Google, Micron, Oracle, and Goldman Sachs.

Introduction

We apply the TensorFlow end-to-end open source platform within the Google Colaboratory (Colab) ecosystem to demonstrate state-of-the-art deep neural network models with hands-on Python code exercises for intermediate to advanced Python users. The Colab ecosystem is a product from Google Research that allows anybody to write and execute arbitrary Python code through a browser. The ecosystem is especially suited to deep learning, data analytics, research, and machine learning education applications. The Colab ecosystem is a hosted Jupyter notebook service that requires no setup to use while providing free access to powerful computing resources such as Graphics Processing Units (GPUs) and Tensor Processing Units (TPUs).

The book is organized into 14 chapters. Chapter 1 introduces you to TensorFlow input pipelines with the tf.data application programming interface (API), which enables you to build complex input pipelines from simple, reusable pieces. Input pipelines are the lifeblood of any deep learning experiment because learning models expect data in a TensorFlow consumable form. Chapter 2 leverages input pipelines to create augmented data experiments that increase the diversity of a training set by applying random (but realistic) transformations. Chapter 3 introduces TensorFlow Datasets to you. TensorFlow offers over 255 practice datasets that are preprocessed for consumption by deep learning models. These datasets are created by the Google Brain TensorFlow team to provide a diverse set of data for practicing machine and deep learning experiments. Chapter 4 demonstrates two end-to-end deep learning experiments with large and complex TensorFlow dataset (TFDS) objects. Chapter 5 introduces TPUs with code examples. A Tensor Processing Unit is an application-specific integrated circuit (ASIC) designed to accelerate machine learning workloads. Chapter 6 introduces transfer learning. Transfer learning is the process of creating new learning models by fine-tuning previously trained neural networks. Chapter 7 leverages the transfer learning knowledge from Chapter 6 to work on advanced transfer learning experiments. Chapter 8 shifts focus from supervised learning to unsupervised learning algorithms beginning with simple stacked autoencoders. Chapter 9 introduces advanced autoencoders including convolutional neural network autoencoders and variational autoencoders. Chapter 10 introduces basic generative modeling with a generative adversarial network (GAN) experiment. Chapter 11 dives into more advanced generative modeling with a Progressive Growing GAN experiment. Chapter 12 shifts to computer visioning with a neural style transfer experiment. Chapter 13 delves into more advanced computer visioning with an object detection experiment. Chapter 14 ends the book with a very simple reinforcement learning experiment.

CHAPTER 1

Build TensorFlow Input Pipelines

We introduce you to TensorFlow input pipelines with the tf.data API, which enables you to build complex input pipelines from simple, reusable pieces. Input pipelines are the lifeblood of any deep learning experiment because learning models expect data in a TensorFlow consumable form. It is very easy to create high-performance pipelines with the tf.data.Dataset abstraction (a component of the tf.data API) because it represents a sequence of elements from a dataset in a simple format.

Although data cleaning is a critical component of input pipelining, we focus on building pipelines with cleansed data. We want to focus you on building TensorFlow consumable pipelines rather than data cleansing. A data scientist can spend upwards of 80% of the total machine learning (ML) project's time on just cleaning the data.

We build input pipelines from three data sources. The first data source is from data loaded into memory. The second one is from external files. The final one is from cloud storage.

Notebooks for chapters are located at the following URL:

https://github.com/paperd/deep-learning-models

What Are Input Pipelines?

A **machine learning (ML) input pipeline** is an approach to codify and automate the workflows required to produce a machine learning model. *ML workflows* are the phases that are implemented during a ML project. Typical phases include data collection, data preprocessing, building datasets, model training and refinement, evaluation, and deployment to production. So the goal of an input pipeline is to automate the workflows

D. Paper, *State-of-the-Art Deep Learning Models in TensorFlow*, https://doi.org/10.1007/978-1-4842-7341-8_1

(or phases) associated with ML problem solving. Once an input pipeline is automated, it can be reused as new data is added to a ML project. It can even be tweaked for use with similar ML projects.

The first step in any input pipeline is data preprocessing. In this step, raw data is gathered, cleansed, and merged into a single organized framework. **Data cleaning** is the process of identifying and fixing any issues with a dataset. The objective of data cleaning is to fix any data that is incorrect, inaccurate, incomplete, incorrectly formatted, duplicated, or irrelevant to the purpose of the ML project so that the cleansed dataset is correct, consistent, reliable, and usable.

Without robust and accurate data as input to train a model, projects are more likely to fail. Once data is properly prepared, the focus of an input pipeline is on writing and executing ML algorithms to obtain a ML model.

Why Build Input Pipelines?

To understand the importance of input pipelines, it is beneficial to look at the typical stages that data science teams work through when building a ML model. Implementing a ML model from scratch tends to be very problem-oriented. So a data science team focuses on producing a model to solve a single business problem.

Manual Workflow

Typically, teams start with a *manual workflow* with no existing infrastructure. Data collection, data cleaning, model training, and evaluation are likely written in a single notebook. The notebook is run locally to produce a model, which is handed over to an engineer tasked with turning it into an application programming interface (API) endpoint. An **API endpoint** is a remote tool utilizing ML to solve a specific problem within a specific project. So the engineer works with a trained model to create an API tool that can be deployed across platforms.

A manual workflow is often ad hoc and starts to break down when a team begins to speed up its iteration cycle because manual processes are difficult to repeat and document. Code in a single notebook format tends to be unsuitable for collaboration. In a manual workflow scenario, the **model** is the product.

Automated Workflow

Once teams move from a stage where they are occasionally updating a single model to having multiple frequently updated models in production, a pipeline approach becomes paramount. In this scenario, we don't build and maintain a model. We develop and maintain a pipeline. So the **pipeline** is the product.

An automated pipeline consists of components and a blueprint for how the components are integrated to produce and update the most crucial component - the model. With automated workflows, code is split into more manageable components including data preprocessing, model training, model evaluation, and retraining triggers. Such triggers are put in place to automatically fire when a model requires retraining.

The system offers the ability to execute, iterate, and monitor a single component in the context of the entire pipeline with the same ease and rapid iteration as running a local notebook cell on a laptop. It also lets us define the required inputs and outputs, library dependencies, and monitored metrics.

The ability to split problem solving into reproducible, predefined, and executable components forces the team to adhere to a joined (or joint) process. A joined process, in turn, creates a well-defined language between the data scientists and engineers that eventually leads to an automated setup that is the ML equivalent of continuous integration (CI). CI is the practice of automating the integration of code changes from multiple contributors into a single software project so the final product is capable of auto-updating itself.

Basic Input Pipeline Mechanics

The *tf.data.Dataset* API supports writing descriptive and efficient input pipelines, which follows a common pattern. First, create a source dataset from the input data. Second, apply data transformations to preprocess the data. Third, iterate over the dataset and process the elements. Iteration happens in a streaming fashion so the entire dataset doesn't need to fit into memory.

Once a data source is created, transform it into a new dataset by chaining method calls on the tf.data.Dataset object. The dataset object is a Python iterable that can be consumed with a for loop.

A TensorFlow dataset is typically created in two distinct ways. We can create a dataset from data stored in memory or in one or more files. If need be, however, we can create a dataset with a data transformation based on one or more tf.data.Dataset objects.

To create a TensorFlow dataset from memory, use either the *from_tensors()* or *from_tensor_slices()* method. To create a TensorFlow dataset from data stored in a file in the recommended TFRecord format, use the *TFRecordDataset()* method.

The **from_tensors()** method combines the input tensor from the data source and returns a dataset with a single element. The **from_tensor_slices()** method creates a dataset with a separate element for each row of the input tensor. An **input tensor** is a vector or matrix of n dimensions that represents the data source. We focus on the *from_tensor_slices()* method because we want to be able to conveniently inspect and process each element from the data source.

High-Performance Pipelines

The tf.data API enables creation of flexible and efficient input pipelines by delivering data for the next step of training before the current step has finished. We focus on three of the best practices for building performant TensorFlow input pipelines, namely, prefetch, cache, and shuffle. We discuss each of these practices with examples when we build input pipelines later in the chapter.

Google Developers Codelabs

Even after you work through the examples in this book, you may want to add to your deep learning application knowledge by exploring additional tutorials. *Google Developers Codelabs* provide guided tutorials emphasizing hands-on coding examples. Most tutorials step you through the process of building a small application or adding a new feature to an existing application. They cover a wide range of topics such as Android Wear, Google Compute Engine, Project Tango, and Google APIs on iOS.

To peruse the Codelabs website, visit
https://codelabs.developers.google.com/

Create a New Notebook in Colab

Within the Colab environment, it is easy to create a new notebook. Open Google Colab in a browser (if not already open). From the pop-up window, click *New notebook*. If already in the Colab environment, click *File* in the top-left menu under *Welcome to*

Colaboratory. Click *New notebook* from the drop-down menu. A code cell is now ready for executing Python code! Add code or text cells by clicking the + *Code* or + *Text* button. For more options, click *Insert* from the main menu.

For an introduction to Colab, peruse

`https://colab.research.google.com/`

To create your first piece of code, add the following in the code cell:

```
10 * 5
```

To execute the code, click the *little arrow to the left*. The output from the code cell shows the result of the multiplication.

Tip We recommend copying and pasting code from the website.

Import the TensorFlow Library

Before we can do anything in TensorFlow, we must import the appropriate Python library. It is common practice to alias the TensorFlow library as **tf**. So go ahead and execute the import in a new code cell:

```
import tensorflow as tf
```

GPU Hardware Accelerator

To vastly speed up processing, use the GPU available from the Google Colab cloud service. Colab provides a free Tesla K80 GPU of about 12 GB RAM (as of this writing). It's very easy to enable the GPU in a Colab notebook:

1. Click *Runtime* in the top-left menu.
2. Click *Change runtime type* from the drop-down menu.
3. Choose *GPU* from the *Hardware accelerator* drop-down menu.
4. Click *Save*.

Note The GPU must be enabled in *each* notebook. But it only has to be enabled once.

Verify that the GPU is active:

```
tf.__version__, tf.test.gpu_device_name()
```

If '/device:GPU:0' is displayed, the GPU is active. If ' ' is displayed, the regular CPU is active.

Tip If you get the error **NAME 'TF' IS NOT DEFINED**, re-execute the code to import the TensorFlow library! For some reason, we sometimes have to re-execute the TensorFlow library import in Colab. We don't know why this is the case.

Colab is a wonderful tool to work with TensorFlow. However, it does have its limitations. Colab applies dynamic resource provisioning. In order to be able to offer computational resources for free, Colab dynamically adjusts usage limits and hardware availability on the fly. So available resources in Colab vary over time to accommodate fluctuations in demand. In a nutshell, this means that Colab may not always be available for use! One solution is to move to Colab Pro for a small monthly fee. As of this writing, the cost is $9.99/month.

Tip For serious TensorFlow users, we recommend moving to Colab Pro. It is not free, but it is quite inexpensive. From our experience, it is more powerful than the free version, and it is more readily available.

Create a TensorFlow Dataset

Create a dataset of three tensors with six elements each:

```
data = [[8, 5, 7, 3, 9, 1],
        [0, 3, 1, 8, 5, 7],
        [9, 9, 9, 0, 0, 7]]

dataset = tf.data.Dataset.from_tensor_slices(data)
dataset
```

Create the dataset. Transform it into a tf.data.Dataset object with the from_tensor_slices() method. The shape of the dataset is (6,), which means that each row contains six scalar values.

Tip We highly recommend testing small pieces of code in their own code cells to reduce debugging time and effort.

Consume the Dataset

Iterate over the dataset to display tensor information:

```
for i, row in enumerate(dataset):
  print ('row ' + str(i), ':', end=' ')
  print (row.numpy())
```

Since the tf.data.Dataset object is created with from_tensor_slices(), it is a Python iterable that can be consumed with a for loop. With TensorFlow datasets, use the *numpy()* method to explicitly convert each tensor to a NumPy array.

Alternatively, we can use the *take()* method to iterate over a TensorFlow dataset:

```
for i, e in enumerate(dataset.take(3)):
  print ('row ' + str(i), ':', end=' ')
  print (e.numpy())
```

We add *3* as a parameter in the take() method to grab three examples.

Another option is to create a Python iterator:

```
i = 0
it = iter(dataset)
print ('row ' + str(i), ':', end=' ')
print (next(it).numpy())
i += 1
print ('row ' + str(i), ':', end=' ')
print (next(it).numpy())
i += 1
print ('row ' + str(i), ':', end=' ')
print (next(it).numpy())
```

Initialize a counter variable. Use the *iter()* method to create an iterator. Consume the iterator with the *next()* method and display the results.

Dataset Structure

The *element_spec* property of tf.data.Dataset allows inspection of the dataset. The property returns a nested structure of the tf.TypeSpec object that matches the structure of the element. The nested structure may be a single component, a tuple of components, or a nested tuple of components.

Inspect the dataset:

```
dataset.element_spec
```

Shape and datatype are displayed.

Alternatively, we can just display the tf.data.Dataset object:

```
dataset
```

Create a Dataset from Memory

If all of your input data fits in memory, the simplest way to create a TensorFlow dataset is to convert it to tf.Tensor objects with the from_tensor_slices() method. Now, we are going to build a pipeline. We begin by loading a clean dataset. We continue by scaling the feature data images. Scaling (or feature scaling) is a method used to normalize the range of independent variables or features of a dataset. Scaling is important because ML models tend to work better if the pixels that make up each image are smaller in size. We inspect the data with code and visualizations. Next, we configure the pipeline for performance. We end by creating a model, training the model, and evaluating the model.

Load and Inspect Data

To build an input pipeline, we need a dataset. Since the focus is on building a TensorFlow consumable pipeline, we work with cleansed datasets.

Load training and test data in memory:

```
train, test = tf.keras.datasets.fashion_mnist.load_data()
```

Download Fashion-MNIST data into training and test sets. We use training data to teach the model. We use test data to evaluate the model. Fashion-MNIST is a dataset of Zalondo's article images. It contains 60,000 training and 10,000 test examples. The dataset is intended to serve as a direct drop-in replacement of the original MNIST dataset for benchmarking machine learning algorithms.

Inspect:

```
type(train[0]), type(train[1])
```

Training and test sets are tuples where the first tuple element contains feature images and the second contains corresponding labels. Both datasets are NumPy arrays.

Load images and labels into variables:

```
train_img, train_lbl = train
test_img, test_lbl = test
```

By separating images and labels from the respective datasets, we can more easily process images and labels as needed.

Verify shapes:

```
print ('train:', train_img.shape, train_lbl.shape)
print ('test:', test_img.shape, test_lbl.shape)
```

Training data consists of 60,000 28 × 28 feature images and 60,000 labels. Test data consists of 10,000 28 × 28 feature images and 10,000 labels.

Scale and Create the tf.data.Dataset

Scale data for efficient processing and create the training and test sets:

```
train_image = train_img / 255.0
test_image = test_img / 255.0

train_ds = tf.data.Dataset.from_tensor_slices(
    (train_image, train_lbl))
test_ds = tf.data.Dataset.from_tensor_slices(
    (test_image, test_lbl))
```

Get slices of the NumPy arrays in the form of tf.data.Dataset objects with *from_tensor_slices()*. Feature image pixel values are typically integers that range from 0 to 255. To scale, divide feature images by 255 to get pixel values that range from 0 to 1.

Scaling images is a critical preprocessing step because deep learning models train faster on smaller images. Moreover, many deep learning model architectures require that images are the same size. But raw images tend to vary in size.

Inspect training and test tensors:

```
train_ds, test_ds
```

Both datasets are *TensorSliceDataset* objects, which means that they are iterators. An **iterator** is an object that contains a countable number of examples that can be traversed with the *next()* method.

Display the first label from the training set:

```
next(train_ds.as_numpy_iterator())[1]
```

Each example in the training set contains an image matrix and its corresponding label. The *next()* method returns a tuple with the first image matrix and its label in positions 0 and 1 in the tuple respectively.

Display ten labels from the training set:

```
next(train_ds.batch(10).as_numpy_iterator())[1]
```

The *batch()* method takes *n* examples from a dataset.

Display all 60,000 labels from the training set:

```
labels = next(train_ds.batch(60_000).as_numpy_iterator())[1]
labels, len(labels)
```

Display the first image from the training set:

```
next(train_ds.as_numpy_iterator())[0]
```

Verify that the first image is a 28 × 28 matrix:

```
arrays = len(next(train_ds.as_numpy_iterator())[0])
pixels = len(next(train_ds.as_numpy_iterator())[0][0])
arrays, pixels
```

To find dimensions of a matrix in Python, the height (or rows) is *len(matrix),* and the width (or columns) is *len(matrix[0]).*

Verify Scaling

Display a pre-scaled tensor from the training set:

```
train_img[0][3]
```

Display the same tensor after scaling:

```
train_image[0][3]
```

Voilà! The pixels are scaled between 0 and 1.

Check Tensor Shape

Check shapes:

```
for img, lbl in train_ds.take(5):
  print ('image shape:', img.shape, end=' ')
  print ('label:', lbl.numpy())
```

Fashion-MNIST images are equally sized. So we don't have to resize them!

Inspect Tensors

Check train and test tensors:

```
train_ds, test_ds
```

All is well.

Preserve the Shape

Assign a variable to the feature image shape for use in the model:

```
for img, _ in train_ds.take(1):
  img.shape
```

```
img_shape = img.shape
img_shape
```

Visualize

Visualize an element from the training set:

```
import matplotlib.pyplot as plt

for feature, label in train_ds.take(1):
  plt.imshow(feature, cmap='ocean')
plt.axis('off')
plt.grid(b=None)
```

Although Fashion-MNIST images are grayscale, we can bring them to life with colors using predefined color maps built into the matplotlib library. A **color map** is an array of colors used to map pixel data to actual color values.

Peruse the following URL for detailed information about matplotlib color maps: *https://matplotlib.org/3.1.0/tutorials/colors/colormaps.html*

Define Class Labels

From our experience working with Fashion-MNIST, we know the corresponding labels:

```
class_labels = ['T-shirt/top', 'Trouser', 'Pullover', 'Dress',
                'Coat', 'Sandal', 'Shirt', 'Sneaker', 'Bag',
                'Ankle boot']
```

Convert a Numerical Label to a Class Label

Labels are numerical in the tf.data.Dataset that we just loaded, but we can display the corresponding class name with the *class_labels* list we just created:

```
for _, label in train_ds.take(1):
  print ('numerical label:', label.numpy())
print ('string label:', class_labels[label.numpy()])
```

Take an example and display the label as a numerical value and string value.

Create a Plot of Examples from the Dataset

Take some images and labels from the training set:

```
num = 30
images, labels = [], []
for feature, label in train_ds.take(num):
  images.append(tf.squeeze(feature.numpy()))
  labels.append(label.numpy())
```

Create a function to display a grid of examples as shown in Listing 1-1.

Listing 1-1. Function to Display a Grid of Examples

```
def display_grid(feature, target, n_rows, n_cols, cl):
  plt.figure(figsize=(n_cols * 1.5, n_rows * 1.5))
  for row in range(n_rows):
    for col in range(n_cols):
      index = n_cols * row + col
      plt.subplot(n_rows, n_cols, index + 1)
      plt.imshow(feature[index], cmap='twilight',
                 interpolation='nearest')
      plt.axis('off')
      plt.title(cl[target[index]], fontsize=12)
  plt.subplots_adjust(wspace=0.2, hspace=0.5)
```

Invoke the function:

```
rows, cols = 5, 6
display_grid(images, labels, rows, cols, class_labels)
```

It's always a good idea to check out the dataset to see if it is as we expect.

Build the Consumable Input Pipeline

We say *consumable input pipeline* because the actual pipeline starts when data is actually acquired. We use this terminology to emphasize the importance of transforming the training and test datasets into efficient tensors for TensorFlow model consumption.

We see examples that refer to this part as building the input pipeline, but the input pipeline encompasses the entire workflow from raw data to generalized model. In later chapters, we drop the word "consumable."

Configure the Dataset for Performance

Use buffered prefetching and caching to improve I/O performance. Shuffle data to improve model performance.

Prefetching is a function in the tf.data API that overlaps data preprocessing and model execution while training, which reduces the overall training time of a model. To perform this operation, add the *tf.Dataset.prefetch* transformation to the input pipeline.

Add the *tf.data.Dataset.cache* transformation to the pipeline to keep images in memory after they're loaded off disk during the first epoch, which ensures that the dataset doesn't become a bottleneck during training. So caching saves operations (e.g., file opening, data reading) from being executed during each epoch.

Shuffling data serves the purposes of reducing variance (ensuring that a model remains general) and reducing overfitting. An obvious case for shuffling is when data is sorted by class (or target). We shuffle to ensure that the training, test, and validation sets are representative of the overall distribution of the data. To perform this operation, add the *tf.Dataset.shuffle* transformation to the pipeline.

Training is always performed on batches of training data and labels. Doing so helps the algorithm converge. **Batch** is when all of a dataset is used to compute the gradient during one iteration. **Mini-batch** is when a subset of a dataset is used to compute the gradient during one iteration. To perform this operation, add the *tf.Dataset.batch* transformation to the pipeline.

The *batch dimension* is typically the first dimension of data tensors. So a tensor of shape [100, 192, 192, 3] contains 100 images of 192 × 192 pixels with three values per pixel (RGB) in each batch. The **RGB color model** is an additive color model in which red, green, and blue lights are added together in various ways to reproduce a broad array of colors.

Build the consumable input pipeline:

```
BATCH_SIZE = 128
SHUFFLE_SIZE = 5000

train_f = train_ds.shuffle(SHUFFLE_SIZE).batch(BATCH_SIZE)
train_fm = train_f.cache().prefetch(1)
```

```
test_f = test_ds.batch(BATCH_SIZE)
test_fm = test_f.cache().prefetch(1)
```

Shuffle training data. Shuffling randomizes training data, which ensures that each data element is independent from other data elements during each training epoch. Learning models tend to perform best when exposed to independently sampled data.

Batch, cache, and prefetch training and test data. Adding the *cache()* transformation increases performance because data is read and written only once during the first epoch rather than during every epoch. Adding the *prefetch(1)* transformation is a good idea because it adds efficiency to the batching process. That is, while our training algorithm is working on one batch, TensorFlow is working on the dataset in parallel to get the next batch ready. So this transformation can dramatically improve training performance.

Like other tf.data.Dataset methods, prefetch operates on the elements of the input dataset. It has no concept of examples vs. batches. So prefetch two examples with examples.prefetch(2) and prefetch two batches with 20 examples per batch with examples.batch(20).prefetch(2).

The test (or validation) set is used to demonstrate how well the trained model works on examples it hasn't seen during training. So it being shuffled is irrelevant.

We set batch size and shuffle size based on trial and error experiments. You can experiment by adjusting batch and shuffle sizes.

Inspect tensors:

```
train_fm, test_fm
```

Build the Model

Import requisite libraries:

```
from tensorflow.keras.models import Sequential
from tensorflow.keras.layers import Dense, Flatten, Dropout
from tensorflow.keras.losses import SparseCategoricalCrossentropy
import numpy as np
```

Clear previous models and generate a seed for reproducibility of results:

```
tf.keras.backend.clear_session()
np.random.seed(0)
tf.random.set_seed(0)
```

We use zero for the seed value, but any number can be substituted.

Tip Clearing previous models does *not* reset the current model to its initial state. To reset a model, just rebuild the input pipeline for the model!

Create the model:

```
model = Sequential([
  Flatten(input_shape=img_shape),
  Dense(128, activation='relu'),
  Dropout(0.4),
  Dense(10, activation=None)
])
```

The basic building block of a neural network is the layer. Layers extract representations from the data fed into them. Hopefully, these representations are meaningful for the problem at hand. Most of deep learning consists of chaining together simple layers. Most layers, such as *Dense,* have parameters that are learned during training.

The first layer in this network is a *Flatten* layer, which transforms the format of the images from a two-dimensional array (of 28 by 28 pixels) to a one-dimensional array (of 28 * 28 = 784 pixels). Think of this layer as unstacking rows of pixels in the image and lining them up. This layer has no parameters to learn because it only reformats the data.

After the pixels are flattened, the network consists of a sequence of two Dense layers. Dense layers are fully connected neural layers, which means that all the neurons in a layer are connected to all neurons in the next layer.

The first Dense layer has 128 nodes (or neurons). We add a *Dropout* layer after the first Dense layer to reduce overfitting. The second (and last) layer returns a logits array with length of 10. **Logits** are the outputs of a layer of neurons before the activation function is applied. Each node contains a score that indicates that the current image belongs to one of the ten classes.

Dropout is a regularization method that approximates training a large number of neural networks with different architectures in parallel. During training, some number of layer outputs are randomly ignored or “dropped out,” which has the effect of making the layer look like and be treated like a layer with a different number of nodes and connectivity to the prior layer. In effect, each update to a layer during training is performed with a different "view" of the configured layer.

Inspect the model:

```
model.summary()
```

Compile and Train the Model

Compile the model with *SparseCategoricalCrossentropy* loss. Sparse categorical cross-entropy performs well when classes are mutually exclusive. That is, each sample belongs exactly to one class. An advantage of using sparse categorical cross-entropy is that it saves time in memory as well as computation because it uses a single integer for a class rather than a whole vector.

The *from_logits=True* attribute informs the loss function that the output values generated by the model are not normalized. That is, the softmax function *has not* been applied on them to produce a probability distribution.

Compile:

```
model.compile(optimizer='adam',
  loss=SparseCategoricalCrossentropy(from_logits=True),
  metrics=['accuracy'])
```

Train the model:

```
epochs = 10
history = model.fit(train_fm, epochs=epochs,
                    verbose=1, validation_data=test_fm)
```

The model is training with ten epochs. The number of **epochs** is the number of times the learning algorithm works through the entire training dataset. Loss and accuracy are displayed for training and test data. Training loss and accuracy are based on what the model learned during training. Test loss and accuracy are based on new data that the model hasn't learned upon. So the closer test accuracy is to training accuracy, the more generalizable the model. Of course, we want to have high test accuracy and low test loss.

Load a TensorFlow Dataset as NumPy

The previous section modeled Fashion-MNIST data based on a Keras dataset. However, we can load data as a TensorFlow dataset (TFDS) and convert it into NumPy arrays for very easy processing. We cover TFDS in detail in a later chapter.

For this experiment, we load the MNIST dataset instead of Fashion-MNIST. We do this because we work with Fashion-MNIST many times in later chapters. So we just want to expose you to another dataset for practice. Once the data is loaded and converted to NumPy, the input pipeline phases are the same as in the previous section.

Create a training set as NumPy arrays in a single batch:

```
import tensorflow_datasets as tfds

image_train, label_train = tfds.as_numpy(
    tfds.load(
        'mnist', split='train',
        batch_size=-1, as_supervised=True,
        try_gcs=True))

type(image_train), image_train.shape
```

By using *batch_size=-1*, the full dataset is loaded as a single batch. The *tfds.load()* function loads the dataset. The *tfds.as_numpy()* function converts the dataset to NumPy arrays.

The training set contains 60,000 28 × 28 images. The *1* dimension indicates that the data is grayscale. A **grayscale** image is one in which the only colors are shades of gray. That is, the image only contains luminance (or brightness) information and no color information.

Create the corresponding test set:

```
image_test, label_test = tfds.as_numpy(
    tfds.load(
        'mnist', split='test',
        batch_size=-1, as_supervised=True,
        try_gcs=True))

type(image_test), image_test.shape
```

Inspect Shapes and Pixel Intensity

Get training shapes:

```
image_train.shape, label_train.shape
```

Get test shapes:

```
image_test.shape, label_test.shape
```

Create a function to find the first pixel vector with pixel intensity values as shown in Listing 1-2.

Listing 1-2. Function to Find Pixel Intensity

```
def find_intensity(m):
  for i, vector in enumerate(m):
    for j, pixels in enumerate(vector):
      if pixels > 0:
        print (vector)
        return i, j
```

Invoke the function:

```
M = image_train[0]
indx = find_intensity(M)
image_train[0][indx[0]][indx[1]]
```

The nonzero values are pixel intensities.

Display the first pixel with intensity greater than zero:

```
image_train[0][indx[0]][indx[1]]
```

Scale

Since NumPy array values are float, divide them by 255 to scale the image pixels:

```
train_sc = image_train / 255.0
test_sc = image_test / 255.0
```

Verify that scaling worked:

```
image_train[0][indx[0]][indx[1]], train_sc[0][indx[0]][indx[1]]
```

Prepare Data for TensorFlow Consumption

Slice NumPy arrays into TensorFlow datasets:

```
train_mnds = tf.data.Dataset.from_tensor_slices(
    (image_train, label_train))
test_mnds = tf.data.Dataset.from_tensor_slices(
    (image_test, label_test))
```

Inspect:

```
train_mnds, test_mnds
```

Build the Consumable Input Pipeline

Initialize parameters, shuffle training data, and batch and prefetch training and test data:

```
BATCH_SIZE = 100
SHUFFLE_SIZE = 10000

train_mnist = train_mnds.shuffle(SHUFFLE_SIZE).\
                         batch(BATCH_SIZE).prefetch(1)
test_mnist = train_mnds.batch(BATCH_SIZE).prefetch(1)
```

Inspect tensors:

```
train_mnist, test_mnist
```

Build the Model

Earlier, we imported requisite libraries. Since they are already in memory, we don't need to import them again (assuming that we are using the same notebook).

Get tensor shape:

```
np_shape = image_test.shape[1:]
np_shape
```

Clear previous models and generate a seed for reproducibility of results:

```
np.random.seed(0)
tf.random.set_seed(0)
tf.keras.backend.clear_session()
```

Create the model:

```
model = Sequential([
  Flatten(input_shape=np_shape),
  Dense(512, activation='relu'),
  Dense(10, activation='softmax')
])
```

Compile and Train the Model

Compile with sparse categorical cross-entropy. Notice that we **don't** set *from_logits=True* because we use *softmax* activation in the output layer of the model to produce a probability distribution from the logits. The **softmax** activation function acts on a vector to increase the difference between the largest component and all others and normalizes the vector to have a sum of 1 so that it can be interpreted as a vector of probabilities. It is used as the last step in classifiers:

```
model.compile(optimizer='adam',
              loss='sparse_categorical_crossentropy',
              metrics=['accuracy'])
```

Train the model:

```
epochs = 3
history = model.fit(train_mnist, epochs=epochs, verbose=1,
                    validation_data=test_mnist)
```

We train for just three epochs because MNIST is so easy to train.

Create a Dataset from Files

Create a TensorFlow dataset from files with a Keras utility. We use the Keras utility because it greatly simplifies processing of the flowers dataset. The *flowers dataset* is public with thousands of flower photos distributed into five classes. Just like the Fashion-MNIST and MNIST examples, we build an input pipeline following similar workflow phases.

Download and Inspect the Dataset

Import a library for visualization:

```
import PIL.Image
```

The dataset contains several thousand photos of flowers in five subdirectories with one flower photo per class. The directory structure is as follows:

```
flowers_photos/
  daisy/
  dandelion/
  roses/
  sunflowers/
  tulips/
```

Download data with the tf.keras.utils.get_file utility:

```
import pathlib

url1 = 'https://storage.googleapis.com/download.tensorflow.org/'
url2 = 'example_images/flower_photos.tgz'
dataset_url = url1 + url2

data_dir = tf.keras.utils.get_file(origin=dataset_url,
                                   fname='flower_photos',
                                   untar=True)
data_dir = pathlib.Path(data_dir)
```

The *tf.keras.utils.get_file* utility downloads a file from a URL if not already in the cache. The *pathlib.Path* function provides a concrete path to the files.

Count the number of flower photos downloaded and available in data_dir:

```
image_count = len(list(data_dir.glob('*/*.jpg')))
print (image_count)
```

There are 3670 files of flower images.

The *data_dir* path points to directories that each hold a different type of flower. Let's see the directories:

```
dirs = [item.name for item in data_dir.glob('*')\
        if item.name != 'LICENSE.txt']
dirs
```

Each directory contains images of that type of flower.

Access some of the files:

```
files = tf.data.Dataset.list_files(str(data_dir/'*/*'))
fn = []
for f in files.take(4):
  print(f.numpy()), fn.append(str(f.numpy()))
```

Display labels from each file:

```
from pathlib import Path

label = []
for i in range(4):
  parts = Path(fn[i]).parts
  label.append(parts[5])
  print (parts[5])
```

Each directory contains images of that type of flower. Here is the first flower in the *daisy* directory:

```
daisy = list(data_dir.glob('daisy/*'))
parts = Path(daisy[0]).parts
print (parts[5])
PIL.Image.open(str(daisy[0]))
```

Display the number of daisy images:

```
len(daisy)
```

Let's display several images from the *roses* directory. Create a list to hold roses:

```
roses = list(data_dir.glob('roses/*'))
```

Grab labels from some of the files:

```
label = []
for i in range(4):
  tup = Path(str(roses[i])).parts
  label.append(tup[5])
```

Display some roses as shown in Listing 1-3.

Listing 1-3. Plot Rose Images

```
rows, cols = 2, 2
plt.figure(figsize=(10, 10))
for i in range(rows*cols):
  plt.subplot(rows, cols, i + 1)
  pix = np.array(PIL.Image.open(str(roses[i])))
  plt.imshow(pix)
  plt.title(label[i])
  plt.axis('off')
```

Notice that images are not of the same size!

Parse Data with the tf.keras.preprocessing Utility

The *tf.keras.preprocessing.image_dataset_from_directory* utility offers incredible convenience for loading and parsing images off disk! We show the convenience of the utility in the "*Create Training and Test Sets*" subsection.

Set Parameters

Set batch size, image height, and image width:

```
BATCH_SIZE = 32
img_height = 180
img_width = 180
```

We set batch size initially to 32 because it tends to be a good size for many of the datasets with which we work. We set image height and width at 180 because we get good results and the model trains really fast. Feel free to experiment with these parameters.

Our inspection revealed that image size differs. Since TensorFlow models expect images of the same size, we must resize them.

Create Training and Test Sets

The *tf.keras.preprocessing.image_dataset_from_directory* utility generates a tf.data. Dataset from image files in a directory. The utility is very useful because it allows us to conveniently split, seed, resize, and batch data. Let's split data into 81% training and 19% test sets. We set this split based on numerous experiments. Of course, you can tweak the sizes with your own experiment. The combination of the *validation_split* and *subset* parameters determines the train and test splits.

Set aside 81% for training data:

```
train_ds = tf.keras.preprocessing.image_dataset_from_directory(
  data_dir,
  validation_split=0.19,
  subset='training',
  seed=0,
  image_size=(img_height, img_width),
  batch_size=BATCH_SIZE)
```

Set aside 19% for test data:

```
test_ds = tf.keras.preprocessing.image_dataset_from_directory(
  data_dir,
  validation_split=0.19,
  subset='validation',
  seed=0,
  image_size=(img_height, img_width),
  batch_size=BATCH_SIZE)
```

Inspect Tensors

Inspect:

```
train_ds, test_ds
```

Take the first batch from the training set and preserve shapes:

```
for img, lbl in train_ds.take(1):
  print (img.shape, lbl.shape)
flower_shape, just_img = img.shape[1:],\
                         img.shape[1:3]
```

We take the first batch to help us inspect a batch from the dataset. We preserve the shape of the batch and batch size for use in the model. Batch size is 32, and images are resized to 180 × 180 × 3. The *3* value indicates that images have three channels, which means they are RGB (color). Labels have shape (32,) that corresponds to the 32 scalar images.

Get Class Names

We already identified the classes from the directory names. But we can now access them with the *class_names* method:

```
class_names = train_ds.class_names
class_names
```

Display Examples

Take a batch from the training set and plot some images as shown in Listing 1-4.

Listing 1-4. Plot Flower Images from the First Training Batch

```
plt.figure(figsize=(10, 10))
for images, labels in train_ds.take(1):
  for i in range(9):
    ax = plt.subplot(3, 3, i + 1)
    plt.imshow(images[i].numpy().astype('uint8'))
    plt.title(class_names[labels[i]])
    plt.axis('off')
```

Scale the Data

As noted earlier in the chapter, a pixel is represented by 256 values. So RGB channel values are in the [0, 255] range. Since neural networks work better with small values, data is typically scaled to be in the [0, 1] range.

Create a function to scale images:

```
def format_image(image, label):
  image = tf.image.resize(image, just_img) / 255.0
  return image, label
```

The function is used when we configure the input pipeline.

Configure the Dataset for Performance

Use buffered prefetching to get data from disk to mitigate I/O issues. Cache data to keep images in memory after they're loaded off disk. Caching saves operations like file opening and data reading from being executed during each epoch.

Build the Input Pipeline

Scale, shuffle the training set, and cache and prefetch train and test sets:

```
SHUFFLE_SIZE = 100

train_fds = train_ds.map(format_image).\
  shuffle(SHUFFLE_SIZE).cache().prefetch(1)
test_fds = test_ds.map(format_image).\
  cache().prefetch(1)
```

Note Since training and test data have already been batched by the utility, *do not* batch when building the input pipeline!

Build the Model

Since we are working with large color images, we need to build a convolutional neural network (CNN) model to garner respectable performance because flower images are color with higher pixel counts.

We need additional libraries for a CNN:

```
from tensorflow.keras.layers import Conv2D, MaxPooling2D
```

Get the number of classes for use in the model:

```
num_classes = len(class_names)
num_classes
```

Clear any previous models and generate a random seed:

```
tf.keras.backend.clear_session()
np.random.seed(0)
tf.random.set_seed(0)
```

Create a multilayer CNN as shown in Listing 1-5.

Listing 1-5. A Multilayer CNN

```
flower_model = tf.keras.Sequential([
  Conv2D(32, 3, activation='relu',
         input_shape=flower_shape),
  MaxPooling2D(),
  Conv2D(32, 3, activation='relu'),
  MaxPooling2D(),
  Conv2D(32, 3, activation='relu'),
  MaxPooling2D(),
  Flatten(),
  Dense(128, activation='relu'),
  Dense(num_classes, activation='softmax')
])
```

The first layer scales the data. The second layer contains 32 neurons with a 3 × 3 convolutional kernel (or filter). Activation is *relu*. The third layer uses maximum pooling to reduce the spatial size of a layer by just keeping the maximum values. As such, the pooling layer reduces image dimensionality without losing important features or patterns. The next four layers repeat the same pattern as the second and third layers. The Flatten layer converts pooled data into a single column because a Dense layer expects data in this form. The final Dense layer enables classification and prediction.

Compile and Train the Model

Compile with SparseCategoricalCrossentropy():

```
flower_model.compile(
  optimizer='adam',
  loss=tf.losses.SparseCategoricalCrossentropy(),
  metrics=['accuracy'])
```

Since *softmax* is applied to outputs, we **don't** set *from_logits=True*.

Train the model:

```
history = flower_model.fit(
    train_fds,
    validation_data=test_fds,
    epochs=5)
```

The model is overfitting because validation accuracy is low compared to training accuracy. But we have not made any attempt to tune the model. In the next chapter, we explore a powerful technique to mitigate overfitting.

Get Flowers from Google Cloud Storage

We demonstrated input pipelining with data from memory and from files. We can also pipeline data from cloud storage. Flowers data is hosted in a public bucket on Google Cloud Storage (GCS). So we can grab flower files from GCS. We can read flowers as JPEG files or as TFRecord files. For data modeling, we use TFRecord files. For optimal performance, we read from multiple TFRecord files at once. The **TFRecord format** is a simple format for storing a sequence of binary records. A TFRecord file contains a sequence of records, which can only be read sequentially.

Read Flowers as JPEG Files and Perform Simple Processing

Read JPEG files based on a GCS pattern:

```
GCS_PATTERN = 'gs://flowers-public/*/*.jpg'
filenames = tf.io.gfile.glob(GCS_PATTERN)
```

GCS_PATTERN is a *glob pattern* that supports the "*" and "?" wildcards. **Globs** (also known as glob patterns) are patterns that can expand a wildcard pattern into a list of pathnames that match the given pattern.

Get the number of JPEG images:

```
num_images = len(filenames)
print ('Pattern matches {} images.'.format(num_images))
```

Create a dataset of filenames from GCS_PATTERN and peruse its contents:

```
filenames_ds = tf.data.Dataset.list_files(GCS_PATTERN)
for filename in filenames_ds.take(5):
  print (filename.numpy().decode('utf-8'))
```

We need the data in (image, label) tuples to work with the images and labels independently. So create a function to return a dataset of (image, label) tuples as shown in Listing 1-6.

Listing 1-6. Function That Returns a Dataset of (image, label) Tuples

```
def decode_jpeg_and_label(filename):
  bits = tf.io.read_file(filename)
  image = tf.image.decode_jpeg(bits)
  label = tf.strings.split(
      tf.expand_dims(filename, axis=-1), sep='/')
  label = label.values[-2]
  return image, label
```

Map the function to each filename to create a dataset of (image, label) tuples:

```
ds = filenames_ds.map(decode_jpeg_and_label)
```

Peruse:

```
for image, label in ds.take(5):
  print (image.numpy().shape,
         label.numpy().decode('utf-8'))
```

Display an image:

```
for img, lbl  in ds.take(1):
  plt.axis('off')
  plt.title(lbl.numpy().decode('utf-8'))
  fig = plt.imshow(img)
```

Although we don't train with this dataset, let's see how to convert text labels to encoded labels as shown in Listing 1-7.

Listing 1-7. Convert Text Labels to Encoded Labels

```
for img, lbl  in ds.take(1):
  label = lbl.numpy().decode('utf-8')

matches = tf.stack([tf.equal(label, s)\
                    for s in class_names], axis=-1)
one_hot = tf.cast(matches, tf.float32)
print (matches.numpy(), one_hot.numpy())
new_label = tf.math.argmax(one_hot)
new_label.numpy()
```

Take a label. Compare it against the class name list to find its position in the list. Create a one-hot vector. Convert the one-hot vector into a label tensor. We don't train with this dataset because it is not the way to model complex data from GCS. But it is a simple way to load and inspect the data.

Read and Process Flowers as TFRecord Files

The best way to model complex data from GCS is as TFRecord files. A TFRecord file stores data as a sequence of binary strings. Binary strings are very efficient to process.

Read TFRecord Files

Read TFRecord files based on a GCS pattern:

```
piece1 = 'gs://flowers-public/'
piece2 = 'tfrecords-jpeg-192x192-2/*.tfrec'
TFR_GCS_PATTERN = piece1 + piece2
tfr_filenames = tf.io.gfile.glob(TFR_GCS_PATTERN)
```

Get the number of buckets:

```
num_images = len(tfr_filenames)
print ('Pattern matches {} image buckets.'.format(num_images))
```

We grabbed 16 buckets. Since there are 3670 flower files, 15 buckets contain 230 images ($15 \times 230 = 3{,}450$), and the final bucket contains 220 images. Add 3,450 to 220 to get 3,670.

Display a file:

```
filenames_tfrds = tf.data.Dataset.list_files(TFR_GCS_PATTERN)
for filename in filenames_tfrds.take(1):
  print (filename.numpy())
```

Set Parameters for Training

Set parameters for image resizing, pipelining, and number of epochs:

```
IMAGE_SIZE = [192, 192]
AUTO = tf.data.experimental.AUTOTUNE
BATCH_SIZE = 64
SHUFFLE_SIZE = 100
EPOCHS = 5
```

Use *AUTOTUNE* to prompt the tf.data runtime, which tunes the value dynamically at runtime.

Note AUTOTUNE is experimental, which means that the operation may change in the future.

Set validation split and class labels:

```
VALIDATION_SPLIT = 0.19
CLASSES = ['daisy', 'dandelion', 'roses', 'sunflowers', 'tulips']
```

Create data splits, validation steps, and steps per epoch as shown in Listing 1-8.

Listing 1-8. Create Training Splits and Steps

```
split = int(len(tfr_filenames) * VALIDATION_SPLIT)
training_filenames = tfr_filenames[split:]
validation_filenames = tfr_filenames[:split]
print ('Splitting dataset into {} training files and {}'
       ' validation files'.format(
           len(tfr_filenames), len(training_filenames),
           len(validation_filenames)), end = ' ')
print ('with a batch size of {}.'.format(BATCH_SIZE))
validation_steps = int(3670 // len(tfr_filenames) *\
                       len(validation_filenames)) // BATCH_SIZE
steps_per_epoch = int(3670 // len(tfr_filenames) *\
                      len(training_filenames)) // BATCH_SIZE
print ('There are {} batches per training epoch and {} '\
      'batches per validation run.'\
      .format(BATCH_SIZE, steps_per_epoch, validation_steps))
```

Create Functions to Load and Process TFRecord Files

Create a function to parse a TFRecord file as shown in Listing 1-9.

Listing 1-9. Function to Parse a TFRecord

```
def read_tfrecord(example):
  features = {
      'image': tf.io.FixedLenFeature([], tf.string),
      'class': tf.io.FixedLenFeature([], tf.int64)
  }
  example = tf.io.parse_single_example(example, features)
  image = tf.image.decode_jpeg(example['image'], channels=3)
  image = tf.cast(image, tf.float32) / 255.0
  image = tf.reshape(image, [*IMAGE_SIZE, 3])
  class_label = example['class']
  return image, class_label
```

The function accepts a TFRecord file. A dictionary holds datatypes common to TFRecords. The tf.string datatype converts the image to byte strings (list of bytes). The tf.int64 converts the class label to a 64-bit integer scalar value. The TFRecord file is parsed into (image, label) tuples. The image element, a JPEG-encoded image, is decoded into a uint8 image tensor. The image tensor is scaled to the [0, 1] range for faster training. It is then reshaped to a standard size for model consumption. The class label element is cast to a scalar.

Create a function to load TFRecord files as tf.data.Dataset as shown in Listing 1-10.

Listing 1-10. Function to Create a tf.data.Dataset from TFRecord Files

```
def load_dataset(filenames):
  option_no_order = tf.data.Options()
  option_no_order.experimental_deterministic = False
  dataset = tf.data.TFRecordDataset(
      filenames, num_parallel_reads=AUTO)
  dataset = dataset.with_options(option_no_order)
  dataset = dataset.map(read_tfrecord, num_parallel_calls=AUTO)
  return dataset
```

The function accepts TFRecord files. For optimal performance, code is included to read from multiple TFRecord files at once. The options setting allows order-altering optimizations. As such, *n* files are read in parallel, and data order is disregarded in favor of reading speed.

Create a function to build an input pipeline from TFRecord files as shown in Listing 1-11.

Listing 1-11. Function to Build an Input Pipeline from TFRecord Files

```
def get_batched_dataset(filenames, train=False):
  dataset = load_dataset(filenames)
  dataset = dataset.cache()
  if train:
    dataset = dataset.repeat()
    dataset = dataset.shuffle(SHUFFLE_SIZE)
  dataset = dataset.batch(BATCH_SIZE)
  dataset = dataset.prefetch(AUTO)
  return dataset
```

The function accepts TFRecord files and calls the *load_dataset* function. The function continues by building an input pipeline by caching, repeating, shuffling, batching, and prefetching the dataset. Repeating and shuffling are only mapped to training data to follow best practices for a Keras dataset.

Create Training and Test Sets

Instantiate the datasets:

```
training_dataset = get_batched_dataset(
    training_filenames, train=True)
validation_dataset = get_batched_dataset(
    validation_filenames, train=False)
training_dataset, validation_dataset
```

Display an image and preserve the shape for the model:

```
for img, lbl in training_dataset.take(1):
  plt.axis('off')
  plt.title(CLASSES[lbl[0].numpy()])
  fig = plt.imshow(img[0])
  tfr_flower_shape = img.shape[1:]
```

Model Data

Clear and seed:

```
tf.keras.backend.clear_session()
np.random.seed(0)
tf.random.set_seed(0)
```

Create the model as shown in Listing 1-12.

Listing 1-12. Create the Model

```
tfr_model = Sequential([
  Conv2D(32, (3, 3), activation = 'relu',
         input_shape=tfr_flower_shape),
  MaxPooling2D(2, 2),
  Conv2D(64, (3, 3), activation='relu'),
```

```
    MaxPooling2D(2, 2),
    Conv2D(128, (3, 3), activation='relu'),
    MaxPooling2D(2),
    Conv2D(128, (3, 3), activation='relu'),
    MaxPooling2D(2, 2),
    Flatten(),
    Dense(512, activation='relu'),
    Dense(num_classes, activation='sigmoid')
])
```

Inspect:

```
tfr_model.summary()
```

Compile:

```
loss = tf.keras.losses.SparseCategoricalCrossentropy(
    from_logits=True)

tfr_model.compile(optimizer='adam',
              loss=loss,
              metrics=['accuracy'])
```

Train:

```
history = tfr_model.fit(training_dataset, epochs=EPOCHS,
                    verbose=1, steps_per_epoch=steps_per_epoch,
                    validation_steps=validation_steps,
                    validation_data=validation_dataset)
```

Summary

We built ML input pipeline examples from three types of data. The first experiments built pipelines from data loaded into memory. We then built a pipeline from external files. The final experiment built a pipeline from cloud storage.

CHAPTER 2

Increase the Diversity of Your Dataset with Data Augmentation

We guide you in the creation of augmented data experiments to increase the diversity of a training set by applying random (but realistic) transformations. Data augmentation is very useful for small datasets because deep learning models crave a lot of data to perform well.

Notebooks for chapters are located at the following URL:

https://github.com/paperd/deep-learning-models

Data Augmentation

More data typically increases model performance. So what do we do if we have small amounts of image training data and cannot collect more? One popular method is data augmentation.

Data augmentation is the process of increasing the size and diversity of an existing training set without manually collecting any new data. The process generates additional training data from existing examples by augmenting them using random transformations that yield believable-looking images. With augmented training data, a learning model is exposed to more aspects of the data, which helps it generalize better. Typically, data augmentation is needed when training data is complex and contains few examples.

Data augmentation is realized by performing a series of random preprocessing transformations to existing data such as horizontal and vertical flipping, skewing, cropping, shearing, zooming in and out, and rotating. Collectively, augmented data is

D. Paper, *State-of-the-Art Deep Learning Models in TensorFlow*, https://doi.org/10.1007/978-1-4842-7341-8_2

able to simulate a variety of subtly different data points as opposed to just duplicating the same data. The subtle differences of the augmented images should (hopefully) be enough to help train a more robust model.

Data augmentation can also mitigate overfitting. Overfitting generally occurs when there are a small number of training examples. By generating additional training data from existing examples, overfitting may be mitigated.

Overfitting happens when a model learns the detail and noise in the training data to the extent that it negatively impacts the performance of the model on new data. So noise or random fluctuations in the training data are picked up and learned as concepts by the model. Small datasets don't contain enough diversity (or randomness) to mitigate learning noise or random fluctuations.

The goal of deep learning is to tune a learning model's parameters so it can effectively map a particular input (e.g., an image) to some output (e.g., a label). Essentially, we try to find the sweet spot where the model's loss is at a minimum, which happens when parameters are tuned in the right way. Data augmentation is an effective tuning mechanism that can help meet the goal of deep learning, especially with small datasets!

For an excellent introduction to data augmentation, peruse *www.tensorflow.org/tutorials/images/data_augmentation*

Import the TensorFlow Library

Import the library and alias it as **tf**:

```
import tensorflow as tf
```

Aliasing the TensorFlow library as tf is common practice.

GPU Hardware Accelerator

Remember from Chapter 1 that you can vastly speed up processing by using the GPU available from the Google Colab cloud service. To save you the trouble of flipping back to Chapter 1, we repeat the instructions to enable the GPU:

1. Click *Runtime* in the top-left menu.
2. Click *Change runtime type* from the drop-down menu.

3. Choose *GPU* from the *Hardware accelerator* drop-down menu.
4. Click *Save*.

Verify that the GPU is active:

```
tf.__version__, tf.test.gpu_device_name()
```

If '/device:GPU:0' is displayed, the GPU is active. If '..' is displayed, the regular CPU is active.

Note If you get the error **NAME 'TF' IS NOT DEFINED**, re-execute the code to import the TensorFlow library!

Augment with a Keras API

In the previous chapter, we experienced overfitting when modeling the flowers dataset. The issue in that chapter is that validation accuracy was low with respect to training accuracy. Let's see if we can mitigate that issue using a Keras API for data augmentation.

Get Data

First, you'll want to get the data again. Do that by executing the following code, which retrieves the same flowers data as was used in Chapter 1:

```
import pathlib

url1 = 'https://storage.googleapis.com/download.tensorflow.org/'
url2 = 'example_images/flower_photos.tgz'
dataset_url = url1 + url2

data_dir = tf.keras.utils.get_file(origin=dataset_url,
                                   fname='flower_photos',
                                   untar=True)
data_dir = pathlib.Path(data_dir)
```

Split Data

The *tf.keras.preprocessing.image_dataset_from_directory* utility reads the data from a directory. The utility is also able to split, seed, resize, and batch the data.

Set the parameters, and as a starting point, try the following three values:

```
BATCH_SIZE = 32
img_height = 180
img_width = 180
```

Batch size is set to 32. We set batch size to 32, but feel free to experiment with different sizes. Since images have different shapes, we must resize them for model consumption. We chose 180 × 180, but feel free to experiment with different image sizes. But be sure to keep height and width the same.

Note Setting batch and image sizes is not a science. The task requires experimentation. From our research, a good starting point for batch size is 32 or 64. Any value smaller than 32 slows learning. Any value larger than 64 is computationally expensive. The idea is to fit a batch of data entirely in memory. Since computer memory comes with a storage capacity in the power of two, it is recommended to keep batch size a power of two. Smaller image sizes speed learning, but preserve less of the original image. So model performance is negatively impacted. Larger image sizes preserve more of the original image. So model performance is positively impacted at the expense of computational resources.

Load Images Off Disk into Train and Test Sets

Now create the train and test datasets. It's common practice to split data into these two set types. Begin with the training set and create it as follows:

```
train_ds = tf.keras.preprocessing.image_dataset_from_directory(
  data_dir,
  validation_split=0.19,
  subset='training',
```

```
  seed=0,
  image_size=(img_height, img_width),
  batch_size=BATCH_SIZE)
```

Set *validation_split=0.19* and *subset='training'* to get 81% of the data into the train set. Set *image size* to the sizes we establish to resize images. Also set the seed and batch values.

Create the test set:

```
test_ds = tf.keras.preprocessing.image_dataset_from_directory(
  data_dir,
  validation_split=0.19,
  subset='validation',
  seed=0,
  image_size=(img_height, img_width),
  batch_size=BATCH_SIZE)
```

Set *validation_split=0.19* and *subset='validation'* to get 19% of the data into the test set. Set *image size* to the sizes we establish to resize images. Also set the seed and batch values.

Inspect the Training Set

Take an example batch and display image shape and label:

```
for images, labels in train_ds.take(1):
  print ('image shape:', images.shape)
  print ('labels:', labels.numpy())
  print ('number of labels in a batch:', len(labels))
```

As expected, we have 32 180 × 180 × 3 images in the first batch. The *3* value means that images are RGB color. We also have 32 corresponding labels. We have 32 images and 32 corresponding labels because we set batch size to 32.

Get Number of Classes

Use the *class_names* method from the utility to display the number of class labels:

```
class_names = train_ds.class_names
num = len(class_names)
num
```

Create a Scaling Function

Create a function to scale feature images:

```
def scale(image, label):
  image = tf.cast(image, tf.float32)
  image /= 255.0
  return image, label
```

Build the Input Pipeline

Scale training and test data. Learning models train faster on smaller images. An input image that is twice as large requires a network to learn from four times as many pixels - and that extra training time adds up. Only shuffle training data. Cache and prefetch training and test data to improve model performance:

```
SHUFFLE_SIZE = 100

train_fds = train_ds.map(scale).shuffle(SHUFFLE_SIZE).\
            cache().prefetch(1)
test_fds = test_ds.map(scale).cache().prefetch(1)
```

Note Since training and test data have already been batched by the utility, *do not* batch when building the input pipeline!

Data Augmentation with Preprocessing Layers

Apply data augmentation with experimental Keras preprocessing layers. The *Keras preprocessing layers API* allows developers to build Keras-native input processing pipelines. So TensorFlow models can accept raw images or raw structured data as input, handle feature normalization, and feature value indexing on their own. Simply, the Keras API makes it much easier to input, preprocess, and model raw data.

We begin by importing requisite libraries and create a simple augmentation model. We can include the preprocessing layers inside a model like other layers.

Let's augment images by randomly flipping horizontally, rotating, zooming, and translating as shown in Listing 2-1.

Listing 2-1. Data Augmentation with Several Preprocessing Layers

```
from tensorflow.keras.layers.experimental.preprocessing\
  import RandomFlip
from tensorflow.keras.layers.experimental.preprocessing\
  import RandomRotation
from tensorflow.keras.layers.experimental.preprocessing\
  import RandomZoom
from tensorflow.keras.layers.experimental.preprocessing\
  import RandomTranslation

data_augmentation = tf.keras.Sequential(
  [
   RandomFlip('horizontal'),
   RandomRotation(0.1),
   RandomZoom(0.1),
   RandomTranslation(height_factor=0.2, width_factor=0.2)
  ]
)
```

Note Keras preprocessing layers are experimental, which means that the operation may change in the future.

We flip images left to right, randomly rotate by a factor of 0.1 (or 10%), randomly zoom by a factor of 0.1 (or 10%), and randomly translate by varying height and width by factors of 0.2 (or 20%), respectively. The choice of layers and parameter value settings was conducted through trial and error experimentations. So feel free to experiment with other augmentations.

For more information on Keras experimental preprocessing layers, peruse

www.tensorflow.org/api_docs/python/tf/keras/layers/experimental/preprocessing

Display an Augmented Image

Here is what happens when applying data augmentation to the same image several times as shown in Listing 2-2.

Listing 2-2. Visualize an Augmented Image

```
import matplotlib.pyplot as plt

plt.figure(figsize=(10, 10))
for images, _ in train_fds.take(1):
  for i in range(9):
    augmented_images = data_augmentation(images)
    ax = plt.subplot(3, 3, i + 1)
    plt.imshow(augmented_images[0])
    plt.axis('off')
```

We display augmentation on the first image with index of 0. Change the index to see other images being augmented. But keep the index value between 0 and 31 to account for the batch size of 32.

Create the Model

Clear previous models and generate a seed for reproducibility:

```
import numpy as np
tf.keras.backend.clear_session()
np.random.seed(0)
tf.random.set_seed(0)
```

Import requisite libraries for data modeling:

```
from tensorflow.keras.models import Sequential
from tensorflow.keras.layers import Dense, Flatten, Dropout
from tensorflow.keras.losses import SparseCategoricalCrossentropy
from tensorflow.keras.layers import Conv2D, MaxPooling2D
```

Build a multilayered CNN as shown in Listing 2-3.

Listing 2-3. Multilayered CNN

```
model = tf.keras.Sequential([
  data_augmentation,
  Conv2D(32, 3, activation='relu'),
  MaxPooling2D(),
```

```
  Conv2D(32, 3, activation='relu'),
  MaxPooling2D(),
  Conv2D(32, 3, activation='relu'),
  MaxPooling2D(),
  Flatten(),
  Dense(128, activation='relu'),
  Dropout(0.5),
  Dense(num)
])
```

Notice that the first layer is the data augmentation that we built!

Compile and Train the Model

Compile with SparseCategoricalCrossentropy(from_logits=True):

```
model.compile(
  optimizer='adam',
  loss=SparseCategoricalCrossentropy(from_logits=True),
  metrics=['accuracy'])
```

We set *from_logits=True* in the loss function because we don't use softmax activation to normalize neuron input into the output layer of the model. Softmax turns logits into probabilities by taking the exponents of each output and normalizing each number by the sum of those exponents so the entire output vector adds up to one. So all probabilities should add up to one. **Logits** are the numeric output of the last linear layer of a multi-class classification neural network. Since we don't use softmax, we inform the compiler to output logits.

Train the model:

```
history = model.fit(
    train_fds,
    validation_data=test_fds,
    epochs=10)
```

Overfitting is mitigated compared with what we experienced without augmentation in the previous chapter. That is, test accuracy is more closely aligned with train accuracy.

Visualize Performance

Let's see the impact of data augmentation with a visualization as shown in Listing 2-4.

Listing 2-4. Visualize Training Performance with Augmentation

```
acc = history.history['accuracy']
val_acc = history.history['val_accuracy']

loss = history.history['loss']
val_loss = history.history['val_loss']

epochs_range = range(10)

plt.figure(figsize=(8, 8))
plt.subplot(1, 2, 1)
plt.plot(epochs_range, acc, label='Training Accuracy')
plt.plot(epochs_range, val_acc, label='Validation Accuracy')
plt.legend(loc='lower right')
plt.title('Training and Validation Accuracy')

plt.subplot(1, 2, 2)
plt.plot(epochs_range, loss, label='Training Loss')
plt.plot(epochs_range, val_loss, label='Validation Loss')
plt.legend(loc='upper right')
plt.title('Training and Validation Loss')
plt.show()
```

Notice that overfitting is mitigated to a large extent for the model we trained for ten epochs. So data augmentation improved performance.

Apply Augmentation on Images

In the previous section, we added Keras preprocessing layers to a neural network to augment. An alternative method is to apply data augmentation transformations directly on images and then feed them to a neural network *without* Keras preprocessing layers.

In this section, we begin by demonstrating a variety of transformations that can be performed on an image. We grab a batch of data from the preprocessed tensors created in the previous section and demonstrate each transformation technique. After demonstrating how to apply various transformations on images, we continue by using a few of these techniques on images and training a model with these augmented images.

Let's begin by grabbing an example image from the train set we just created:

```
for batch_images, _ in train_fds.take(1):
  print ('image shape:', batch_images.shape)
```

Since we already have available preprocessed tensors from the previous section, we grab the first batch of these images from the pipelined training set. Now, we have a batch of 32 images with which to play contained in *batch_images*. We begin by showing you how to work with the images in the batch.

Set an Index Variable

Since batch size is 32, set the value of the index between 0 and 31:

```
indx = 0
indx
```

We set the index to 0 to get the first image from the batch. Feel free to change the index value. But keep in mind that the index value must be between 0 and 31!

Set an Image

Set the index to grab the first image from the batch. Change the index value (between 0 and 31) to display different images.

Our image:

```
our_image = batch_images[indx]
```

Show an Example

Visualize the first image:

```
plt.imshow(our_image)
plt.axis('off')
plt.grid(b=None)
```

Create Functions to Show Images

Create a visualization function to show the original image and modified image as shown in Listing 2-5.

Listing 2-5. Function to Visualize Original Image and Modified Image

```
def show(original_img, trans_img):
  f = plt.figure(figsize=(6, 6))
  f.add_subplot(1,2,1)
  plt.imshow(original_img)
  plt.axis('off')
  f.add_subplot(1,2,2)
  plt.imshow(trans_img)
  plt.axis('off')
  plt.show(block=True)
```

Create a visualization function to show several transformations of an image as shown in Listing 2-6.

Listing 2-6. Visualize Transformations of an Image

```
def show_images(img, indx, trans, p1=None, p2=None, b=False):
  plt.figure(figsize=(10, 10))
  for i in range(9):
    ax = plt.subplot(3, 3, i + 1)
    if not b:
      new_img = trans(img[indx])
    elif p2==None:
      new_img = trans(img[indx], p1)
      new_img = np.clip(new_img, 0, 1)
    else:
      new_img = trans(img[indx], p1, p2)
      new_img = np.clip(new_img, 0, 1)
    plt.imshow(new_img)
    plt.axis('off')
```

Crop an Image

To crop an image, remove or adjust its outside edges.

Crop the image:

```
new_image = tf.image.random_crop(our_image, [120, 120, 3])
new_image.shape
```

The operation slices a *shape size* portion out of the image at a uniformly chosen offset. In this case, the new image size is 120 × 120.

Show the original image and modified image:

```
show(our_image, new_image)
```

Since the transformation is random, show multiple images:

```
show_images(batch_images, indx, tf.image.random_crop,
            [120, 120, 3], b=True)
```

We show random crops with this visualization. Notice that each image is cropped a bit differently because the API produces crops randomly.

Centrally crop the image:

```
new_image = tf.image.central_crop(our_image, 0.5)
print (new_image.shape)
show(our_image, new_image)
```

The operation reduces image size by half and centrally crops it. That is, it removes background noise and ensures that the remaining image pixels are centered.

Crop to a bounding box:

```
new_image = tf.image.crop_to_bounding_box(
    our_image, 10, 10, 120, 120)
print (new_image.shape)
show(our_image, new_image)
```

The operation crops an image to a specified bounding box.

A **bounding box** is an area in a pixel image defined by two longitudes and two latitudes. Each latitude is a decimal value between -90.0 and 90.0. Each longitude is a decimal value between -180.0 and 180.0.

Randomly Flip Image Left to Right

Since the transformation is random, the image isn't always flipped left to right:

```
show_images(batch_images, indx, tf.image.random_flip_left_right)
```

Randomly Flip Image Up to Down

Since the transformation is random, the image isn't always flipped up to down:

```
show_images(batch_images, indx, tf.image.random_flip_up_down)
```

Flip Image Up to Down

The image is always flipped up to down:

```
show(our_image, tf.image.flip_up_down(our_image))
```

Rotate Image 90 Degrees

Rotate the image 90° by setting *k=1*:

```
show(our_image, tf.image.rot90(our_image, k=1))
```

The *k* argument is the number of times the image is rotated by 90 degrees.
Rotate the image 180° by setting *k=2*:

```
show(our_image, tf.image.rot90(our_image, k=2))
```

Rotate the image 270° by setting k=3:

```
show(our_image, tf.image.rot90(our_image, k=3))
```

If we set k=4, we're back where we started because it rotates the image 360°!

Adjust Gamma

Gamma encoding is used to optimize the usage of bits when encoding an image by taking advantage of the nonlinear manner that humans perceive light and color. Human perception of brightness (or lightness) under common illumination conditions (neither pitch black nor blindingly bright) follows an approximate power function with greater sensitivity to relative differences between darker tones than between lighter tones. Simply, gamma encoding can be thought of as the intensity of an image.

Adjust the image with gamma encoding:

```
new_image = tf.image.adjust_gamma(
    our_image, gamma=0.75, gain=1.5)
new_image = np.clip(new_image, 0, 1)
show(our_image, new_image)
```

Adjust gamma for brightness. For gamma less than 1, the image is brighter. For gamma greater than 1, the image is darker. Intensity is controlled by the *gain* argument. Experiment with the *gamma* and *gain* parameters to see the impact on the transformed image.

Note Think of the gamma and gain values as knobs that can be adjusted to modify image appearance.

Adjust Contrast

Contrast in image processing is the difference in luminance (or color) that makes an object distinguishable. It is determined by the difference in the color and brightness of the object and other objects within the same field of view.

Fixed contrast:

```
new_image = tf.image.adjust_contrast(
    our_image, contrast_factor=1.8)
new_image = np.clip(new_image, 0, 1)
show(our_image, new_image)
```

Adjust the *contrast_factor* argument for more or less luminance. Experiment with the *contrast_factor* parameter to see its impact on the transformed image.

Random contrast:

```
show_images(batch_images, indx, tf.image.random_contrast,
            0.75, 2.9, b=True)
```

Adjust lower and upper bounds to change the random contrast boundaries.

Adjust Brightness

Fixed brightness:

```
new_image = tf.image.adjust_brightness(our_image, .25)
new_image = np.clip(new_image, 0, 1)
show(our_image, new_image)
```

Adjust the delta value to increase the value of the images' pixels. So a delta of 0.25 adds 25% brightness to the image.

Random brightness:

```
show_images(batch_images, indx, tf.image.random_brightness,
            0.25, b=True)
```

Adjust Saturation

Saturation in image processing is the depth or intensity of color present within an image. The more saturated an image, the more colorful and vibrant it appears. Less color saturation makes an image appear subdued or muted.

Fixed saturation:

```
show(our_image, tf.image.adjust_saturation(our_image, 3.0))
```

Adjust the saturation factor to increase saturation. So a saturation factor of 3.0 triples the saturation of an image.

Random saturation:

```
show_images(batch_images, indx, tf.image.random_saturation,
            0.3, 3.5, b=True)
```

Adjust lower and upper bounds to change the random saturation boundaries.

Hue

Hue is the main indication of a RGB color. It is the value that tells us whether an object is red, green, or blue. In contrast, saturation is the perceived intensity. So saturation is how colorful the object looks, while hue is the actual color.

Fixed hue:

```
show(our_image, tf.image.adjust_hue(our_image, 0.2))
```

Random hue:

```
show_images(batch_images, indx, tf.image.random_hue, 0.2, b=True)
```

Adjust the 0.2 parameter value to see its impact on the transformed image.

To see all possible data augmentation transformations, peruse *www.tensorflow.org/api_docs/python/tf/image/*

For a general site for applying augmentation directly on images, peruse *https://towardsdatascience.com/tensorflow-image-augmentation-on-gpu-bf0eaac4c967*

Apply Transformations Directly on Images

Now that we've covered several potential transformations, let's see if we can improve training performance. We apply just a couple of transformations with pretty good results, but you can experiment with as many as you wish.

Create an Augmentation Function

Create a function that randomly flips an image from left to right and applies a saturation operation to an image as shown in Listing 2-7.

Listing 2-7. Augmentation Function

```
def augment(image, label):
  img = tf.image.random_flip_left_right(image)
  final_image = tf.image.random_saturation(img, 0, 2)
  return (final_image, label)
```

With so many transformation options and little guidance that we could find, it is not easy to find the right mix. We tried many variations, but the preceding simple one worked the best for us. We encourage you to experiment with different transformations to see if you can improve learning.

Display an Augmented Image

Here is what happens when the *augment* function is applied to the same image several times as shown in Listing 2-8.

Listing 2-8. Plot Transformations for an Image

```
plt.figure(figsize=(10, 10))
for i in range(9):
  image, _ = augment(our_image, labels[0])
  ax = plt.subplot(3, 3, i + 1)
  plt.imshow(image)
  plt.axis('off')
```

Build the Input Pipeline

Build the pipeline for train and test data:

```
SHUFFLE_SIZE = 100

train1 = train_ds.map(scale, num_parallel_calls=4)
train2 = train1.map(augment, num_parallel_calls=4)
train_da = train2.shuffle(SHUFFLE_SIZE).cache().prefetch(1)
test_da = test_ds.map(scale).cache().prefetch(1)
```

Notice that we map the augment function only to training data. We add the *num_parallel* parameter for better performance during training.

Note Only augment training images! Training data is presented to a model to help it become more generalizable and robust. Test data is presented as new data to help evaluate a model.

Create the Model

Clear all previous models and generate a seed for reproducibility:

```
tf.keras.backend.clear_session()
np.random.seed(0)
tf.random.set_seed(0)
```

Create a multilayer CNN as shown in Listing 2-9.

Listing 2-9. Multilayer CNN

```
model = tf.keras.Sequential([
  Conv2D(32, 3, activation='relu',
         input_shape=[180, 180, 3]),
  MaxPooling2D(),
  Conv2D(32, 3, activation='relu'),
  MaxPooling2D(),
  Conv2D(32, 3, activation='relu'),
  MaxPooling2D(),
  Flatten(),
  Dense(128, activation='relu'),
  Dropout(0.5),
  Dense(num)
])
```

Notice that the model doesn't have Keras preprocessing layers like the one in the previous section. Since we are directly augmenting images, we don't need to add the Keras layers.

Compile and Train the Model

Compile with SparseCategoricalCrossentropy(from_logits=True):

```
model.compile(
  optimizer='adam',
  loss=SparseCategoricalCrossentropy(from_logits=True),
  metrics=['accuracy'])
```

Train the model:

```
history = model.fit(
    train_da,
    validation_data=test_da,
    epochs=5)
```

Voilà. We mitigated overfitting! We used five epochs, but you can substitute any number of epochs you wish. More epochs require more memory. Since our dataset is not very large, memory is really not an issue. But with large datasets, memory can become an issue if you set the number of epochs really high.

Data Augmentation with ImageGenerator

So far, we have presented two augmentation techniques. The first technique added Keras preprocessing layers to a model. The second one applied transformations directly on images and then fed them to a model. Both techniques built a tf.data pipeline to feed to a model.

An alternative is to use the ImageGenerator class. The *ImageGenerator class* makes it easy to load images from disk and augment them in various ways. We can shift, rotate, rescale, flip horizontally or vertically, shear or apply transformation functions to images. Although ImageGenerator is very convenient for simple projects, building a tf.data pipeline is more conducive to complex projects because it can read images in parallel from any source (not just a local disk) and manipulate the dataset in any manner. Also, preprocessing functions based on tf.image operations can be used in the tf.data pipeline and in the model deployed to production.

We don't advocate one over the other. It depends on the task at hand. The ImageGenerator class is very easy to use. So it should be useful at the beginning stages of a project. The tf.data pipeline is optimized for parallel processing, so it would be a good choice for ongoing projects that require a lot of computing resources.

For a great resource on this subject, peruse

www.tensorflow.org/api_docs/python/tf/keras/preprocessing/image/ImageDataGenerator

Process Flowers Data

The process is simpler than the previous techniques. Load the data directly from the flowers directory into training and test splits by utilizing dictionaries. The first dictionary provides scaling and splitting information to the ImageDataGenerator method within the ImageGenerator class. The second dictionary provides target and batch information to the method.

Import the appropriate library:

```
from tensorflow.keras.preprocessing.image\
  import ImageDataGenerator
```

Create a dictionary to scale and split the data:

```
datagen_kwargs = dict(rescale=1./255, validation_split=.19)
```

Data is split 81% into training and 19% into testing.

Create a dictionary to resize images, set batch size, and interpolate:

```
BATCH_SIZE = 32
IMAGE_SIZE = (180, 180)

dataflow_kwargs = dict(target_size=IMAGE_SIZE,
                       batch_size=BATCH_SIZE,
                       interpolation='bilinear')
```

Create Datasets

Now that the directories are set up correctly, we use a Keras API to create the training and test sets. The test set is created first based on the construction of the first dictionary.

Create the test set:

```
valid_datagen = tf.keras.preprocessing.image.\
  ImageDataGenerator(**datagen_kwargs)
valid_generator = valid_datagen.flow_from_directory(
    data_dir, subset='validation', shuffle=False,
    **dataflow_kwargs)
```

Create the training set:

```
train_datagen = valid_datagen
train_generator = train_datagen.flow_from_directory(
    data_dir, subset='training', shuffle=True,
    **dataflow_kwargs)
```

Notice that we only shuffle training data.

Inspect the tensors:

```
valid_datagen, train_datagen
```

Both tensors are ImageDataGenerator objects. So tensors are ready for training because the class takes care of all preprocessing!

Create the Model

Clear previous models and generate a seed:

```
tf.keras.backend.clear_session()
np.random.seed(0)
tf.random.set_seed(0)
```

Create a multilayer CNN as shown in Listing 2-10.

Listing 2-10. Multilayer CNN for ImageDataGenerator Objects

```
model = Sequential([
  Conv2D(filters=32, kernel_size=(5, 5), activation = 'relu',
         input_shape=(180, 180, 3)),
  MaxPooling2D(2),
  Conv2D(64, (5, 5), activation='relu'),
  MaxPooling2D(2),
  Flatten(),
  Dense(64, activation='relu'),
  Dense(5, activation='softmax')
])
```

We could have used the same model as the one in the previous section, but we used a slightly different one to see what would happen. We used one less convolutional layer and applied softmax to model outputs.

Compile and Train the Model

Compile the model:

```
model.compile(optimizer='adam',
              loss='categorical_crossentropy',
              metrics=['accuracy'])
```

Since *softmax* activation is applied to the neurons coming into the output layer of our model, we **don't** compile with *from_logits=True* for the loss metric. Softmax assigns decimal probabilities to each class in a multi-class problem that must add up to 1.0. Adding this additional constraint can help training converge more quickly than it otherwise would.

Train the model:

```
history = model.fit(train_generator, batch_size=BATCH_SIZE,
                    epochs=5, validation_data=valid_generator,
                    verbose=1)
```

Without data augmentation, overfitting begins to occur in earnest after just a few epochs.

Augment Training Data

Let's see if using ImageDataGenerator to create data augmentations on training data improves performance:

```
aug_train_datagen = tf.keras.preprocessing.\
  image.ImageDataGenerator(
      rotation_range=40, horizontal_flip=True,
      width_shift_range=0.2, height_shift_range=0.2,
      shear_range=0.2, zoom_range=0.2,
      **datagen_kwargs)
```

Rotate and flip images. And shift, shear, and zoom images. Since only training data is used for learning, we don't augment validation (or test) data. We include many parameter values in this example. We did quite a bit of experimentation to arrive at these values. We strongly encourage you to do your own experimentation.

Create the train set with transformations to images:

```
aug_train_generator = aug_train_datagen.flow_from_directory(
    data_dir, subset='training', shuffle=True, **dataflow_kwargs)
```

Inspect:

```
aug_train_generator
```

The tensor is a *DirectoryIterator* object.

Recompile and Train the Model

Since we just augmented the training data, we must retrain the model. So begin by clearing previous model sessions and generating a seed for reproducibility. Continue by recompiling the model.

Clear models and generate a seed:

```
tf.keras.backend.clear_session()
np.random.seed(0)
tf.random.set_seed(0)
```

Recompile the model:

```
model.compile(optimizer='adam',
              loss='categorical_crossentropy',
              metrics=['accuracy'])
```

We recompile the model because we are training on a new dataset.

Train the model:

```
history = model.fit(aug_train_generator, batch_size=BATCH_SIZE,
                    epochs=5, validation_data=valid_generator,
                    verbose=1)
```

The model exhibits less overfitting!

Inspect the Data

ImageDataGenerator data is different than tf.data. So let's explore it. Begin by processing a batch from the original training set and one from the augmented one. Although using the class is pretty easy, it does take a bit more work to explore the data.

Grab a batch from the original training set as shown in Listing 2-11.

Listing 2-11. Process a Batch from the Original Training Set

```
data_list = []
batch_index, end_index = 0, 1
while batch_index <= train_generator.batch_index:
  if batch_index < end_index:
    data = train_generator.next()
    data_list.append(data[0])
    batch_index = batch_index + 1
  else: break
original = np.asarray(data_list)
```

To traverse an ImageDataGenerator object, use the dataset name with a starting index value (assigned to a variable) as the method. Since we just want a single batch, set the ending index value to 1. Grab the first element with the *next()* method. Append it to a list and increment the batch index. Break the loop after one iteration. Finally, convert the list to a NumPy array.

Inspect the shape:

```
original[0].shape
```

As expected, the shape is (32, 180, 180, 3). So each batch consists of 32 180 × 180 × 3 color images.

Verify that we grabbed one batch:

```
print ('We grabbed', len(original), 'batch.')
```

Grab a batch from the augmented training set as shown in Listing 2-12.

Listing 2-12. Process a Batch from the Augmented Training Set

```
data_list = []
batch_index, end_index = 0, 1
while batch_index <= aug_train_generator.batch_index:
  if batch_index < end_index:
    data = aug_train_generator.next()
    data_list.append(data[0])
    batch_index = batch_index + 1
  else: break
augmented = np.asarray(data_list)
```

Inspect the shape:

```
augmented[0].shape
```

As expected, the shape is (32, 180, 180, 3). So each batch consists of 32 180 × 180 × 3 color images.

Verify that we grabbed one batch:

```
print ('We grabbed', len(augmented), 'batch.')
```

Visualize

Grab the first image from the original training set:

```
train_image = original[0][0]
```

Visualize the image:

```
plt.imshow(train_image)
plt.axis('off')
plt.grid(b=None)
```

We see a normal flower image.

Visualize images from the original training set as shown in Listing 2-13.

Listing 2-13. Visualize Several Original Training Images

```
plt.figure(figsize=(10, 10))
for images in original:
  for i in range(9):
    ax = plt.subplot(3, 3, i + 1)
    plt.imshow(images[i])
    plt.axis('off')
```

We see several normal flower images.

Grab the first image from the augmented training set:

```
aug_train_image = augmented[0][0]
```

Visualize the augmented image:

```
plt.imshow(aug_train_image)
plt.axis('off')
plt.grid(b=None)
```

We see an augmented image.

Visualize images from the augmented training set as shown in Listing 2-14.

Listing 2-14. Visualize Several Augmented Training Images

```
plt.figure(figsize=(10, 10))
for images in augmented:
  for i in range(9):
    ax = plt.subplot(3, 3, i + 1)
    plt.imshow(images[i])
    plt.axis('off')
```

We see several augmented flower images. Augmented data looks different than the original data by design because we want to provide the model new data. If we give the data copies of the original data, performance is not enhanced. But augmented data gives the model original data.

Summary

We demonstrated data augmentation with three different techniques. The Keras technique is pretty easy to implement, but doesn't provide as much flexibility as directly applying augmentations on images. The ImageGenerator technique is the easiest to implement, but is limited to smaller projects.

CHAPTER 3

TensorFlow Datasets

We introduce TensorFlow Datasets by discussing and demonstrating their many facets with code examples. Although TensorFlow Datasets are not ML models, we include this chapter because we use them in many of the chapters in this book. These datasets are created by the TensorFlow team to provide a diverse set of data for practicing ML experiments.

When we began working with TensorFlow, we were unfamiliar with the mechanics of TensorFlow Datasets. We worked mainly with NumPy data. So we had to spend quite a bit of time becoming familiar with them. We believe that this chapter should help you navigate these datasets with ease. If you already have experience with TensorFlow Datasets, you may not need to work through the examples in this chapter.

Notebooks for chapters are located at the following URL:

https://github.com/paperd/deep-learning-models

An Introduction to TensorFlow Datasets

TensorFlow Datasets (TFDSs) provide a collection of datasets that are ready to use with TensorFlow or other machine language frameworks (e.g., Jax, Apache Spark, Accord. NET). All TFDSs are exposed as tf.data.Datasets, which allows us to easily build high-performance input pipelines with the tf.data API. It is also very easy to download and prepare TFDSs deterministically, which makes them dataset builders. A **dataset builder** is an object that leverages a set of methods to prepare data for consumption by a ML model.

Don't confuse TFDSs with tf.data. The tf.data API allows us to build efficient data pipelines. TFDSs are a high-level wrapper around tf.data.

For an excellent overview of TFDSs, peruse

www.tensorflow.org/datasets/overview

D. Paper, *State-of-the-Art Deep Learning Models in TensorFlow*, https://doi.org/10.1007/978-1-4842-7341-8_3

Import the TensorFlow Library

Import the library and alias it as **tf**:

```
import tensorflow as tf
```

GPU Hardware Accelerator

For convenience, we repeat the instructions to enable the GPU:

1. Click *Runtime* in the top-left menu.
2. Click *Change runtime type* from the drop-down menu.
3. Choose *GPU* from the *Hardware accelerator* drop-down menu.
4. Click *Save*.

Verify that the GPU is active:

```
tf.__version__, tf.test.gpu_device_name()
```

If '/device:GPU:0' is displayed, the GPU is active. If '..' is displayed, the regular CPU is active.

Note If you get the error **NAME 'TF' IS NOT DEFINED**, re-execute the code to import the TensorFlow library!

Available Datasets

All dataset builders are a subclass of the tfds.core.DatasetBuilder class. So each TFDS object is defined as a tfds.core.DatasetBuilder object. To get the list of available TFDS builders, use tfds.list_builders():

```
import tensorflow_datasets as tfds

tfds.list_builders()
```

Let's see how many are currently available:

```
len(tfds.list_builders())
```

Wow! There are a lot of TFDSs available for deep learning practice!

To find out more about TFDSs, peruse *www.tensorflow.org/datasets/catalog/overview*

Load a Dataset

All builders include a *tfds.core.DatasetInfo* object, which contains a dataset's metadata. Metadata is accessed through either the *tfds.load* API or *tfds.core.DatasetBuilder* API. The tfds.load API is a thin wrapper around tfds.core.DatasetBuilder. So it is much easier to use. From our research, it is the preferred method to download a TFDS. We get the same output using either API.

Load a dataset with tfds.load:

```
ds, info = tfds.load('mnist', split='train',
                     shuffle_files=True,
                     with_info=True,
                     try_gcs=True)
ds
```

The tfds.load API downloads the data and saves it as TFRecord files. It then loads the TFRecord files and creates a tf.data.Dataset. The *TFRecord format* is a simple format for storing a sequence of binary records. So the variable *ds* contains a tf.data.Dataset consisting of MNIST data. The data is shuffled, the meta dataset is returned, and the dataset is retrieved from the Google Cloud Service (GCS). The *MNIST* database is a large database of handwritten digits that is widely used for training and testing in the field of ML.

Common arguments available to the tfds.load API include

split - Splits data (e.g., 'train,' ['train,' 'test'], 'train[80%:],' etc.).

shuffle_files - Controls whether to shuffle the files between each epoch (TFDSs store big datasets in multiple smaller files).

as_supervised - If True, tf.data.Dataset has a two-tuple structure (input, label). If False, tf.data.Dataset has a dictionary structure.

data_dir - Location where the dataset is saved (defaults to ~/ tensorflow_datasets/).

with_info=True - Returns the tfds.core.DatasetInfo containing dataset metadata.

download=False - Disables download.

The *try_gcs* argument is not listed in the documentation. It tells the loader to retrieve the dataset from the GCS.

TFDS Metadata

To access all metadata, just display the contents of the *info* object:

```
info
```

From the metadata, we see that MNIST contains 60,000 28 × 28 feature images for training and 10,000 28 × 28 feature images for testing. Both training and test sets have corresponding scalar labels.

Iterate Over a Dataset

As a Dictionary

By default, a tf.data.Dataset contains a dictionary of tf.Tensors. Let's look at an example from the dataset:

```
ds = ds.take(1)
ds
```

Use *take()* to get *n* examples from a tf.data.Dataset. In this case, we take one example. The example is in the form *{'image': tf.Tensor, 'label': tf.Tensor}*. Simply, each example contains an image tensor and a label tensor. The image tensor is a 28 × 28 matrix, and the label tensor is a scalar.

Image data are datatype tf.uint8, which are 8-bit unsigned integers. A *uint8* datatype contains whole numbers that range from 0 to 255. As with all unsigned numbers, the values must be nonnegative. Uint8s are mostly used in graphics. As a note, colors (pixels that represent colors) are always nonnegative.

Label data are datatype tf.int64, which are 64-bit signed integers. A *int64* datatype contains signed integers that range from negative 9,223,372,036,854,775,808 through positive 9,223,372,036,854,775,807.

Display information about the example:

```
for example in ds:
  print ('keys:', list(example.keys()))
  image = example['image']
  label = example['label']
  print ('shapes:', image.shape, label)
```

A loop is needed to extract image and label information from the dataset.

As Tuples

By using *as_supervised=True,* a tf.data.Dataset contains tuples of features and labels:

```
ds = tfds.load('mnist', split='train', as_supervised=True,
               try_gcs=True)
ds = ds.take(1)

for image, label in ds:
  print (image.shape, label)
```

In the iterate statement, simply include variables to hold features and labels.

As NumPy Arrays

The *tfds.as_numpy()* function converts a tf.data.Dataset to an iterable of NumPy arrays:

```
ds = tfds.load('mnist', split='train', as_supervised=True,
               try_gcs=True)
ds = ds.take(1)

for image, label in tfds.as_numpy(ds):
  print (type(image), type(label), label)
  print (image.shape)
```

We load training data as feature image tuples. We then take one example and convert it to a NumPy array.

Conveniently, we can load the entire dataset as NumPy arrays (if memory permits):

```
image_train, label_train = tfds.as_numpy(
    tfds.load('mnist', split='train',
              batch_size=-1, as_supervised=True,
              try_gcs=True))

type(image_train), image_train.shape
```

By using *batch_size=-1*, the full dataset is loaded in a single batch. The batch is then converted to NumPy arrays.

Note Be careful that your dataset can fit in memory and that all examples have the same shape before training.

Since the dataset consists of NumPy arrays, we can inspect it with normal Python operations. Get the number of examples:

```
len(list(image_train))
```

As expected, there are 60,000 feature image tuples in the training set.

Inspect the first example:

```
image_train[0].shape, label_train[0]
```

The image tensor has a shape of 28 × 28 × 1. The "1" dimension means that the image is grayscale. The label is a scalar that represents the class of the image.

Inspect a few examples:

```
for row in range(3):
  print (image_train[row].shape, label_train[row])
```

Tip If you prefer working with NumPy arrays rather than tf.Tensor objects, load the dataset in a single batch if memory permits.

Visualization

We can conveniently visualize images from a TFDS object.

tfds.as_dataframe

One way to visualize image data is to convert tf.data.Dataset objects to pandas. DataFrame objects. Load the dataset, take four examples, and display:

```
ds, info = tfds.load('mnist', split='train', with_info=True,
                     try_gcs=True)

tfds.as_dataframe(ds.take(4), info)
```

We loaded the dataset with its metadata (*with_info=True*) to enable display.

Take Examples

We can also visualize examples taken from the dataset. Take four examples, squeeze out the "1" dimension from each tensor, and add the squeezed tensors to an array. We need to squeeze out the "1" dimension because the *imshow()* function expects a two-dimensional matrix as input:

```
import matplotlib.pyplot as plt

images = []
for example in ds.take(4):
  img = tf.squeeze(example['image'])
  images.append(img)
```

Visualize the images as shown in Listing 3-1.

Listing 3-1. Visualize Four Images from MNIST

```
rows, cols = 2, 2
plt.figure(figsize=(10, 10))
for i in range(rows*cols):
  plt.subplot(rows, cols, i + 1)
  plt.imshow(images[i], cmap='bone')
  plt.axis('off')
```

tfds.show_examples

Another way to visualize images is to use *show_examples()*:

```
fig = tfds.show_examples(ds, info)
```

Note Only image datasets are supported with show_examples.

Several images are neatly displayed. Underneath each image is the label name as a string and as an integer in parentheses. For example, the image of 4 has 4(4) underneath it because its label is "4" as a string and (4) as an integer.

Load Fashion-MNIST

Load Fashion-MNIST as a TFDS object. *Fashion-MNIST* is a dataset of Zalando's article images consisting of a training set of 60,000 examples and a test set of 10,000 examples. Zalando is an online fashion company that leverages artificial intelligence (AI) to improve the customer experience. The dataset is intended to serve as a direct drop-in replacement of the original MNIST dataset for benchmarking ML algorithms. The original MNIST dataset is being phased out as a practice dataset because it is so easy to achieve stellar performance from a very simple neural network. Fashion-MNIST is recommended because it is much more challenging.

Load the dataset:

```
fashion, fashion_info = tfds.load(
    'fashion_mnist',
    split='train',
    with_info=True,
    shuffle_files=True,
    as_supervised=True,
    try_gcs=True)
```

We load only training data. Metadata is available in the *info* object. We also shuffle the dataset to ensure that each data point creates an *independent change* on the model without being biased by the same points before them. By setting *as_supervised=True,* the returned tf.data.Dataset has a two-tuple structure represented as (input, label). By setting *try_gcs=True*, the dataset is streamed directly from the Google Cloud Service (GCS).

Inspect an example:

```
for image, label in fashion.take(1):
  print (image.shape, label)
```

The tf.Tensor is composed of an image and a label. Image shape is 28 × 28 × 1. The "1" dimension means that the image is grayscale. The label is a scalar value that represents the class of the image.

Metadata

Inspect the dataset:

```
fashion
```

As expected, images in the dataset are 28 × 28 × 1 tensors and labels are scalars.

Display the metadata:

```
fashion_info
```

We see a lot of information about the dataset. A very important one is how the data is split. Conveniently, Fashion-MNIST data is already split into training and test sets. So we don't have to manually split the dataset!

Access the tfds.features.FeatureDict:

```
fashion_info.features
```

The features dictionary contains information about images and labels. Such information is commonly referred to as metadata. The following code snippets provide metadata.

Get the number of classes:

```
num_classes = fashion_info.features['label'].num_classes
num_classes
```

So we have ten class labels. We place the number of classes in a variable for use in the model.

Get class names:

```
classes = fashion_info.features['label'].names
classes
```

We see the names of each class label. Placing class labels in a variable is useful for visualization.

Get shape information:

```
print (fashion_info.features.shape)
print (fashion_info.features.dtype)
print (fashion_info.features['image'].shape)
print (fashion_info.features['image'].dtype)
```

We see shape and datatypes of the features.

Display Split Information

Access tfds.core.SplitDict:

```
fashion_info.splits
```

Get available splits:

```
list(fashion_info.splits.keys())
```

Get available information on the train split:

```
print (fashion_info.splits['train'].num_examples)
print (fashion_info.splits['train'].filenames)
print (fashion_info.splits['train'].num_shards)
```

Visualize

It's a good idea to visualize a dataset before moving forward! Before any deep learning experiment, we explore the metadata and visualize examples to get an idea of what we are going to be working with before conducting any analysis. The idea is to *get to know your data*!

Show some examples:

```
fig = tfds.show_examples(fashion, fashion_info)
```

Underneath each image is the label name as a string and as an integer in parentheses. For example, the image of a coat has Coat(4) underneath it because its label is "Coat" as a string and (4) as an integer.

Show examples from a dataframe:

```
tfds.as_dataframe(fashion.take(4), info)
```

Take some examples and visualize:

```
images, labels = [], []
for image, label in fashion.take(4):
  img = tf.squeeze(image)
  images.append(img), labels.append(label)
```

Visualize the images as shown in Listing 3-2.

Listing 3-2. Visualize Four Images from Fashion-MNIST

```
rows, cols = 2, 2
plt.figure(figsize=(10, 10))
for i in range(rows*cols):
  plt.subplot(rows, cols, i + 1)
  plt.imshow(images[i], cmap='bone')
  t = classes[labels[i]] + ' (' +\
      str(labels[i].numpy()) + ')'
  plt.title(t)
  plt.axis('off')
```

Slicing API

All builder datasets expose various data subsets defined as splits (e.g., [train, test]). When constructing a tf.data.Dataset, we can specify which split(s) we wish to slice. We can also retrieve slice(s) of split(s) as well as combinations of those. In this section, we provide some examples on how to split and slice a TFDS. It is a good idea to know how to manually split a dataset into training and test sets in case you come across data that is not presplit. It is also a good idea to know how to slice a data split (e.g., train, test).

Slicing Instructions

We specify slicing instructions in the tfds.load objects that we create. Instructions are provided either as strings or with the ReadInstruction API. Strings are more compact and readable for simple cases, while the ReadInstruction API provides more options and might be easier to use with variable slicing parameters. Loading data as strings means that the dataset is a Python string. Loading data as a ReadInstruction object means that the dataset is a Python object.

Note Due to the shards being read in parallel, order isn't guaranteed to be consistent between sub-splits. So reading test[0:100] followed by test[100:200] may yield examples in a different order than reading test[:200].

Instructions as Strings

Let's work through some tfds.load examples:

Load the full train set:

```
fashion_train = tfds.load('fashion_mnist', split='train',
                          try_gcs=True)
fashion_train
```

In this case, we load (or slice) the train split from Fashion-MNIST. So slice is just another word for load in this context.

Load the full "train" split and full "test" split as two distinct datasets:

```
train_ds, test_ds = tfds.load('fashion_mnist',
                              split=['train', 'test'],
                              try_gcs=True)
train_ds, test_ds
```

Load the full "train" and "test" splits interleaved together:

```
train_test_ds = tfds.load('fashion_mnist', split='train+test',
                          try_gcs=True)
train_test_ds
```

Load the slice from record 100 (included) to record 200 (excluded) from the "train" split:

```
train_100_200_ds = tfds.load('fashion_mnist',
                             split='train[100:200]',
                             try_gcs=True)
```

In this case, we load a slice from the train set. So slice has a different meaning than load in this context.

Load the slice of the first 25% from the "train" split:

```
train_25pct_ds = tfds.load('fashion_mnist',
                           split='train[:25%]',
                           try_gcs=True)
```

Load the slice from the first 10% of "train" to the last 80% of "train":

```
train_10_80pct_ds = tfds.load(
    'fashion_mnist', try_gcs=True,
    split='train[:10%]+train[-80%:]')
```

Perform tenfold cross-validation as shown in Listing 3-3.

Listing 3-3. Tenfold Cross-Validation

```
test_cv = tfds.load('fashion_mnist', try_gcs=True,
                    split=[f'train[{k}%:{k+10}%]'
                    for k in range(0, 100, 10)])
train_cv = tfds.load('fashion_mnist', try_gcs=True,
                     split=[f'train[:{k}%]+train[{k+10}%:]'
                     for k in range(0, 100, 10)])
```

Cross-validation is a technique to evaluate predictive models by partitioning the original sample into a training set to train the model and a test set to evaluate it. *Tenfold cross-validation* is a common implementation of the technique.

The cross-validation procedure begins by randomly shuffling the dataset. The next step is to split the dataset into *k* groups. For each k group, take the group as a holdout or test dataset. Take the remaining groups as a training dataset. Continue as normal by fitting a model on the training set and evaluating it on the test set. Each observation in the data sample is assigned to an individual group and must stay in that group for the duration of the procedure. So each sample is given the opportunity to be used in the holdout set *1* time and used to train the model *k - 1* times.

For our tenfold cross-validation, here is some documentation:

> * Validation datasets are 10%:
>
> [0%:10%], [10%:20%], ..., [90%:100%]
>
> * Training datasets are the complementary 90%:
>
> [10%:100%] (for a corresponding validation set of [0%:10%])
>
> [0%:10%] + [20%:100%] (for a validation set of [10%:20%])
>
> [0%:90%] (for a validation set of [90%:100%]).

Instructions with the ReadInstruction API

We show the equivalent instructions with the ReadInstruction API starting with loading the full train set:

```
train_ds = tfds.load('fashion_mnist', try_gcs=True,
                     split=tfds.core.ReadInstruction('train'))
```

The same data is loaded, but into an object rather than a string.

Load the full "train" split and full "test" split as two distinct datasets:

```
train_ds, test_ds = tfds.load(
    'fashion_mnist', try_gcs=True,
    split=[tfds.core.ReadInstruction('train'),
           tfds.core.ReadInstruction('test')])
```

Load the full "train" and "test" splits interleaved together:

```
ri = tfds.core.ReadInstruction('train')\
     + tfds.core.ReadInstruction('test')
train_test_ds = tfds.load('fashion_mnist',
                          split=ri, try_gcs=True)
```

Load the slice from record 100 (included) to record 200 (excluded) of the "train" split:

```
train_100_200_ds = tfds.load(
    'fashion_mnist',
    split=tfds.core.ReadInstruction(
        'train', from_=100, to=200,
        unit='abs'), try_gcs=True)
```

Load the slice of the first 25% of the “train” split:

```
train_25_pct_ds = tfds.load(
    'fashion_mnist', try_gcs=True,
    split=tfds.core.ReadInstruction(
        'train', to=25, unit='%'))
```

Load the slice from the first 10% of train to the last 80% of train:

```
ri = (tfds.core.ReadInstruction('train', to=10, unit='%') +
      tfds.core.ReadInstruction('train', from_=-80, unit='%'))
train_10_80pct_ds = tfds.load('fashion_mnist',
                              split=ri, try_gcs=True)
```

Perform tenfold cross-validation as shown in Listing 3-4.

Listing 3-4. Tenfold Cross-Validation with ReadInstruction

```
tests = tfds.load('fashion_mnist', split=
    [tfds.core.ReadInstruction('train', from_=k,
                               to=k+10, unit='%')
     for k in range(0, 100, 10)], try_gcs=True)
trains = tfds.load('fashion_mnist', split=
    [tfds.core.ReadInstruction('train', to=k, unit='%') +
     tfds.core.ReadInstruction('train', from_=k+10, unit='%')
     for k in range(0, 100, 10)], try_gcs=True)
```

For our tenfold cross-validation, here is some documentation:

> * Validation datasets are each going to be 10%: [0%:10%], [10%:20%], ..., [90%:100%].

> * Training datasets are each going to be the complementary 90%:

[10%:100%] (for a corresponding validation set of [0%:10%])

[0%:10%] + [20%:100%] (for a validation set of [10%:20%])

[0%:90%] (for a validation set of [90%:100%])

Performance Tips

This section is intended for those interested in learning more about improving TensorFlow performance. We show three tips, but there are many more. To delve deeper, peruse the following URLs:

www.tensorflow.org/guide/data_performance
www.tensorflow.org/datasets/performances

Auto-caching

By default, TFDSs auto-cache datasets that satisfy the following constraints:

* Total dataset size (all splits) is defined and < 250 MB.

* shuffle_files is disabled or only a single shard is read.

So don't fiddle with auto-caching unless you want to change the default.

Benchmark Datasets

Use *tfds.core.benchmark(ds)* to benchmark any tf.data.Dataset object.

We can load and preprocess a dataset in one step:

```
ds = tfds.load('fashion_mnist', split='train',
               try_gcs=True).batch(32).prefetch(1)
```

Notice that we just grab one batch from the dataset.

Benchmark the dataset:

```
tfds.core.benchmark(ds, batch_size=32)
```

Be sure to initialize the *batch_size* parameter to the same value!

Run the benchmark again:

```
tfds.core.benchmark(ds, batch_size=32)
```

The second iteration benchmark is much faster due to auto-caching!

Reloading a TFDS Object

The first time we load a TFDS object, it takes quite a bit of time especially if the dataset is large. The next time we load a TFDS object, it is much faster because it is already in memory! If you want to see this in action, create a new notebook and load the Fashion-MNIST TFDS object. Load it a second time and notice that it takes very little time.

Load Fashion-MNIST as a Single Tensor

We don't have to load a TFDS as a TensorFlow tensor. We can load it as a NumPy tf.Tensor if the dataset fits in memory. The recommended way is to load the full dataset as a single tensor (or NumPy array). It is possible to do this by setting *batch_size=-1* to batch all examples as a single tf.Tensor.

Load the full dataset as a single tf.Tensor and convert it to a NumPy array:

```
(img_train, label_train), (img_test, label_test) = tfds.as_numpy(
    tfds.load(
        'fashion_mnist', try_gcs=True, as_supervised=True,
        split=['train', 'test'], batch_size=-1))
```

Note To be able to manipulate a tf.Tensor like a NumPy array, load it as a TFDS as a single tensor.

Display shapes:

```
img_train.shape, label_train.shape
```

Train data images have shape (60000, 28, 28, 1) and labels have shape (60000,).

Ready Data for TensorFlow Consumption

Scale feature images and convert data to tf.data.Dataset objects:

```
train = img_train / 255.0
test = img_test / 255.0

train_ds = tf.data.Dataset.from_tensor_slices(
    (train, label_train))
test_ds = tf.data.Dataset.from_tensor_slices(
    (test, label_test))
```

Since the single tensor is a NumPy tensor, we can scale images simply by dividing by 255.0 as we would with a NumPy array.

Build the Input Pipeline

Set parameters and build the input pipeline for train and test data:

```
BATCH_SIZE = 128
SHUFFLE_SIZE = 5000

train_f = train_ds.shuffle(SHUFFLE_SIZE).batch(BATCH_SIZE)
train_fm = train_f.cache().prefetch(1)

test_f = test_ds.batch(BATCH_SIZE)
test_fm = test_f.cache().prefetch(1)
```

Batch size and shuffle size are set based on trial and error experimentation. We recommend that you try different batch sizes (e.g., 32, 64) to see their impacts on learning performance. Shuffle size doesn't seem to impact learning performance as much as batch size. So we recommend that you keep this constant in your experiments.

Build the Model

Get the input shape for the model:

```
img_shape = img_train.shape[1:]
img_shape
```

Import requisite libraries:

```
from tensorflow.keras.models import Sequential
from tensorflow.keras.layers import Dense, Flatten, Dropout
from tensorflow.keras.losses import SparseCategoricalCrossentropy
import numpy as np
```

Clear previous models and generate a seed for reproducibility:

```
tf.keras.backend.clear_session()
np.random.seed(0)
tf.random.set_seed(0)
```

Create the model:

```
model = Sequential([
  Flatten(input_shape=img_shape),
  Dense(128, activation='relu'),
  Dropout(0.4),
  Dense(num_classes, activation=None)
])
```

Compile and Train the Model

Compile with SparseCategoricalCrossentropy(from_logits=True):

```
model.compile(optimizer='adam',
  loss=SparseCategoricalCrossentropy(from_logits=True),
  metrics=['accuracy'])
```

The *from_logits=True* attribute informs the loss function that the output values generated by the model are not normalized. That is, the softmax function has not been applied to the output layer neurons to produce a probability distribution. Unless otherwise specified, softmax activation is automatically applied to the output values. But we explicitly indicate that no activation be applied.

Train the model:

```
epochs = 10
history = model.fit(train_fm, epochs=epochs,
                    verbose=1, validation_data=test_fm)
```

Load Beans as a tf.data.Dataset

Let's load a different dataset as a tf.data.Dataset object rather than as a single tf.Tensor as we did with Fashion-MNIST. Unless you prefer working with NumPy tensors, this is the preferred way to work with a TFDS. We strongly believe that working with various TFDS examples helps you better understand how to work with this type of data. The beans dataset is a good one because it doesn't contain that many examples and there are only three classes.

Beans is a dataset containing images of beans taken in the field using smartphone cameras. It consists of three classes. Two are disease classes and the other is a healthy class. Diseases depicted are angular leaf spot and bean rust. Data was annotated by experts from the National Crops Resources Research Institute (NaCRRI) in Uganda and collected by the Makerere AI research lab.

Examples are presplit into test, training, and validation sets. The test and training sets contain 128 and 1,034 examples, respectively. The validation set contains 133 examples. Of course, you can manually split the dataset differently if you wish.

Load the dataset:

```
beans, beans_info = tfds.load(
    'beans', with_info=True, as_supervised=True,
    try_gcs=True)
```

Inspect the data:

```
beans
```

We see that examples are split into test, training, and validation sets containing 500 × 500 × 3 feature images. So we don't have to split the data ourselves.

For deep learning experiments, datasets are typically split into training and test sets. In industry, it is recommended to split data into training, validation, and test sets. Typically, training data is for learning, validation data is for tuning, and test data is for generalizability. But, we can use test data for tuning and validation data for generalizability.

Splitting data into three sets is superior for industrial purposes because the test set has never been touched by the learning model. So it can be used more confidently for generalizability. Professional data scientists are proficient at tuning models from the validation set. Most online tutorials use only the training and test splits because the focus is on learning rather than application in industry.

Metadata

Inspect the info object:

```
beans_info
```

Inspect shapes:

```
beans['train'], beans['test'], beans['validation']
```

Inspect splits:

```
beans_info.splits
```

From metadata, we can get class labels and number of classes:

```
class_labels = beans_info.features['label'].names
num_classes = beans_info.features['label'].num_classes
class_labels, num_classes
```

Visualize

We can visualize a TFDS in several ways.

Display examples with the show_examples method:

```
fig = tfds.show_examples(beans['train'], beans_info)
```

A TFDS includes a method to display a few examples. Labels are displayed with the class name as a string and numerical value underneath the image.

We can also display examples as a dataframe:

```
tfds.as_dataframe(beans['train'].take(4), info)
```

Finally, we can display examples manually. Begin by building a grid to display multiple examples. Continue by selecting images from the train set:

```
num = 30
images, labels = [], []
for feature, label in beans['train'].take(num):
  images.append(tf.squeeze(feature.numpy()))
  labels.append(label.numpy())
```

Create a function to display the grid of examples as shown in Listing 3-5.

Listing 3-5. Function That Plots a Grid of Examples

```
def display_grid(feature, target, n_rows, n_cols, cl):
  plt.figure(figsize=(n_cols * 1.5, n_rows * 1.5))
  for row in range(n_rows):
    for col in range(n_cols):
      index = n_cols * row + col
      plt.subplot(n_rows, n_cols, index + 1)
      plt.imshow(feature[index], cmap='twilight',
                 interpolation='nearest')
      plt.axis('off')
      t = ' ('  + str(target[index]) + ')'
      plt.title(cl[target[index]] + t, fontsize=7.5)
  plt.subplots_adjust(wspace=0.2, hspace=0.5)
```

Plot the grid:

```
rows, cols = 5, 6
display_grid(images, labels, rows, cols, class_labels)
```

Display the first healthy bean as shown in Listing 3-6.

Listing 3-6. Visualization of the First Healthy Bean

```
for img, lbl in beans['train'].take(30):
  if lbl.numpy() == 2:
    plt.imshow(img)
    plt.axis('off')
    print (class_labels[lbl.numpy()], end=' ')
    print (lbl.numpy())
    break
```

Check Shapes

Although we know that images are all the same shape from the metadata, check manually:

```
for i, example in enumerate(beans['train'].take(5)):
  print('Image {} shape: {} label: {}'.\
        format(i+1, example[0].shape, example[1]))
```

Just like it is a good idea to visualize examples, it is a good idea to manually inspect the shape of a few examples. We display five examples. Just change the number (parameter value) in the take method to vary the number of examples.

Reformat Images

We don't have to resize images because they are all the same shape. However, the images are pretty large. To improve training performance, we resize images to a smaller size.

Create a function to resize and scale images:

```
IMAGE_RES = 224

def format_image(image, label):
  image = tf.image.resize(image, (IMAGE_RES, IMAGE_RES))/255.0
  return image, label
```

We resize images to the same shape because the learning model expects images of the same size. We encourage you to experiment with image size, but don't make the images too big because larger images consume more memory.

Configure Dataset for Performance

It's good practice to use buffered prefetching and caching to improve I/O performance. So we apply both techniques when building the input pipeline.

Prefetching overlaps the preprocessing and model execution of a training step. While the model is executing training step *s*, the input pipeline is reading the data for step *s+1*. Doing so reduces the step time to the maximum (as opposed to the sum) of the training process and the time it takes to extract the data. Apply the *tf.data.Dataset.prefetch* transformation to overlap data preprocessing and model execution during training.

The *tf.data.Dataset.cache* transformation caches a dataset in memory or on local storage. Using this transformation saves some operations (like file opening and data reading) from being executed during each epoch. Specifically, the tf.data.Dataset.cache transformation keeps images in memory after they're loaded off disk during the first epoch. As a result, the dataset doesn't become a bottleneck during training. If the dataset is too large to fit into memory, use this operation to create a performant on-disk cache.

Build the input pipeline:

```
BATCH_SIZE = 32
SHUFFLE_SIZE = 500

train_batches = beans['train'].shuffle(SHUFFLE_SIZE).\
  map(format_image).batch(BATCH_SIZE).cache().prefetch(1)

validation_batches = beans['test'].\
  map(format_image).batch(BATCH_SIZE).cache().prefetch(1)
```

Build the Model

Get input shape:

```
for img, lbl in train_batches.take(1):
  in_shape = img.shape[1:]
in_shape
```

Input shape is TensorShape([224, 224, 3]) as expected. We now have the input shape as a variable to use in the learning model.

Import libraries not already in memory:

```
from tensorflow.keras.layers import Conv2D, MaxPooling2D
```

Clear models and generate a seed:

```
tf.keras.backend.clear_session()
np.random.seed(0)
tf.random.set_seed(0)
```

Create a multilayered CNN as shown in Listing 3-7.

Listing 3-7. Multilayered CNN

```
model = Sequential([
  Conv2D(32, (3, 3), activation = 'relu',
         input_shape=in_shape, strides=1,
         kernel_regularizer='l1_l2'),
  MaxPooling2D(2, 2),
  Conv2D(64, (3, 3), activation='relu'),
  MaxPooling2D(2, 2),
  Conv2D(128, (3, 3), activation='relu'),
  MaxPooling2D(2),
  Conv2D(128, (3, 3), activation='relu'),
  MaxPooling2D(2, 2),
  Flatten(),
  Dense(512, activation='relu'),
  Dense(num_classes, activation='sigmoid')
])
```

Since feedforward networks perform poorly with large images, we need a CNN. The model includes four convolutional layers for two-dimensional spatial data. A **convolution** is the simple application of a filter to an input (an image in our case) that results in an activation. Repeated application of the same filter to an input results in a map of activations called a feature map. A *feature map* indicates the locations and strength of a detected feature in an input such as an image.

Each convolutional layer is a MaxPooling2D layer that performs a max pooling operation for 2D spatial data. **Max pooling** down-samples the input representation from a convolutional layer by taking the maximum value over the window defined by the pool size for each dimension along the features axis. The window is shifted by strides in each dimension.

Pooling involves selecting a pooling operation like a filter to be applied to feature maps. The size of the pooling operation or filter is smaller than the size of the feature map. Specifically, it is almost always 2 × 2 pixels applied with a stride of 2 pixels. A **stride** is the number of pixels shifted over the input matrix (or image). When the stride is 1, we move the filters 1 pixel at a time. When the stride is 2, we move the filters 2 pixels at a time and so on.

Compile and Train the Model

Compile with SparseCategoricalCrossentropy(from_logits=True):

```
loss = tf.keras.losses.\
       SparseCategoricalCrossentropy(from_logits=True)
```

```
model.compile(optimizer='adam',
              loss=loss,
              metrics=['accuracy'])
```

We use this loss function because activation in the output layer of our model is 'sigmoid.' We use this activation because it works well with the dataset. Experiment with different activation functions.

Train the model:

```
epochs = 10
history = model.fit(
    train_batches, epochs=epochs,
    verbose=1, validation_data=validation_batches)
```

Predict

Since we used test data for tuning, make predictions based on the *validation dataset* (because it has never been seen by the model). Build an input pipeline for the validation set to ready it for predictions:

```
validate = beans['validation'].\
  map(format_image).batch(BATCH_SIZE).cache().prefetch(1)
```

Make predictions:

```
predictions = model.predict(validate)
```

The variable contains predictions for each example. Each prediction is an array containing ten predictions. So we need an additional step to get the actual prediction.

Get the actual prediction for the first example:

```
first_prediction = tf.math.argmax(predictions[0])
class_labels[first_prediction.numpy()]
```

Use the tf.math.argmax API to get the actual prediction from the prediction array.

Get multiple predictions as shown in Listing 3-8.

Listing 3-8. Get Multiple Predictions

```
p = []
for row in range(8):
  pred = tf.math.argmax(predictions[row])
  p.append(pred.numpy())
  print ('class:', '(' + str(pred.numpy()) + ')', end=' ')
  print (class_labels[pred.numpy()])
```

We get eight predictions. Change the range parameter value to get as many or few as you wish.

Get an idea of model accuracy:

```
for i, (_, label) in enumerate(beans['validation'].take(8)):
  if label.numpy() == p[i]:
    print ('correct')
  else:
    print ('incorrect', end=' ')
    print ('actual:', label.numpy(), 'predicted:', p[i])
```

For the first eight predictions, we see how well the model performed. We can get overall accuracy by taking the entire validation set and computing the average.

Summary

We learned through examples how to work with TFDSs. We also trained on Fashion-MNIST and beans data for practice with TFDSs.

CHAPTER 4

Deep Learning with TensorFlow Datasets

In the previous chapter, we demonstrated how to work with TFDS objects. In this chapter, we work through two end-to-end deep learning experiments with large and complex TFDS objects. The Fashion-MNIST and beans datasets are small with simple images.

The first experiment works with a large dataset containing complex images of cats and dogs. The second one works with a dataset containing complex images of human hands. The human hands dataset is not as large as the one in the first experiment. Both experiments include data augmentation to enhance performance. The goal is to help you work through end-to-end learning experiments with larger and more complex datasets.

Notebooks for chapters are located at the following URL:

https://github.com/paperd/deep-learning-models

An Experiment with cats_vs_dogs

We demonstrate an end-to-end example with the *cats_vs_dogs* dataset. The dataset contains 23,262 examples of large images of cats and dogs. The goal of this experiment is to provide you experience training a large and complex TFDS. We encourage you to work through many datasets to gain more experience with deep learning. Data is the key! Working with various datasets provides different experiences because each dataset is unique.

Note Working with large image files requires a lot of RAM. So you may need to invest in more memory or subscribe to Google's Colab Pro service if your computer crashes. On our PC, we have had RAM crashes with this dataset on occasion. But since we upgraded to Colab Pro, we haven't had any issues.

D. Paper, *State-of-the-Art Deep Learning Models in TensorFlow*, https://doi.org/10.1007/978-1-4842-7341-8_4

Import the TensorFlow Library

Import the library and alias it as **tf**:

```
import tensorflow as tf
```

Aliasing the TensorFlow library as tf is common practice.

GPU Hardware Accelerator

As a convenience, we provide the steps to enable the GPU in a Colab notebook:

1. Click *Runtime* in the top-left menu.
2. Click *Change runtime type* from the drop-down menu.
3. Choose *GPU* from the *Hardware accelerator* drop-down menu.
4. Click *Save.*

Verify that the GPU is active:

```
tf.__version__, tf.test.gpu_device_name()
```

If '/device:GPU:0' is displayed, the GPU is active. If '..' is displayed, the regular CPU is active.

Note If you get the error **NAME 'TF' IS NOT DEFINED**, re-execute the code to import the TensorFlow library!

Begin the Experiment

When working with a new dataset, it is a good idea to explore its metadata. We can then make an informed decision on how to split it.

Load the TFDS Object

Load the dataset with simple parameters to get its metadata:

```
import tensorflow_datasets as tfds

data, info = tfds.load(name='cats_vs_dogs', with_info=True,
                       try_gcs=True)
```

It takes a bit longer to load the dataset because it contains over 20,000 large images. But once the dataset is loaded into memory, reloading the dataset is very fast. Corrupted images are automatically skipped.

Metadata

Display the contents of the info object:

```
info
```

A lot of information is contained in this object. But we only need to access a few pieces of information from it. Although you may not need a lot of the information contained in the object, it doesn't hurt to display it and read through what is available.

Get class labels and number of classes:

```
class_labels = info.features['label'].names
num_classes = info.features['label'].num_classes
class_labels, num_classes
```

From the metadata, we see that the dataset is *not* presplit. So we must split the data ourselves. We chose to split 80% for the training set, 10% for the validation set, and 10% for the test set. You can choose different splits if you wish.

Display split information:

```
num_train_img = info.splits['train[0%:80%]'].num_examples
num_validation_img = info.splits['train[80%:90%]'].num_examples
num_test_img = info.splits['train[90%:100%]'].num_examples
print ('train images:', num_train_img)
print ('validation images:', num_validation_img)
print ('test images:', num_test_img)
```

Using our split scheme (80:10:10), we should have 18,610 examples for training, 2,326 examples for validation, and 2,326 examples for testing. The motivation for separating data into training, validation, and test splits is to learn from the training data, tune from the validation data, and generalize from the test data.

Note We are not actually splitting the data in this section. We are just showing you how many examples will be in each split with our split scheme.

Although it is common to see data separated into just training and test splits, it is advantageous to add one more split to hold aside some data that the model has never touched. With three splits, training data is for learning, validation data is for tweaking and tuning, and test data is for evaluating for generalizability. With just a training and test split, we must tune and generalize from the test set. Since the model touches the test set, it is not entirely new data!

Verify split percentages:

```
train_num = num_train_img /23262
validation_num = num_validation_img /23262
test_num = num_test_img /23262
```

```
'{0:.0%}'.format(train_num), '{0:.0%}'.format(validation_num),\
'{0:.0%}'.format(test_num)
```

Since the dataset contains 23,262 images, we verify the number of images in each set by dividing by that number.

Split the Data

We can load data into one container and manually perform the desired splits. Or we can just include split information in the split parameter as so:

```
(training_set, validation_set, test_set), info = tfds.load(
    'cats_vs_dogs', with_info=True,
    split=['train[:80%]', 'train[80%:90%]',
           'train[90%:]'], shuffle_files=True,
    as_supervised=True, try_gcs=True)
```

The code loads 80% of the data into the training set, 10% of the data into the validation set, and 10% of the data into the test set in a single step. Previously, we demonstrated how to load the training and test sets in one step.

Manually verify the number of examples in each split:

```
len(list(training_set)), len(list(validation_set)),\
len(list(test_set))
```

The list operation takes some time because each tensor example is processed into a list. So we **don't** recommend performing this operation for REALLY large datasets!

Visualize

Let's visualize some examples in the training set to see what the images and labels look like. We can visualize three ways.

Display examples with show_examples:

```
fig = tfds.show_examples(training_set, info)
```

Display examples as a dataframe:

```
tfds.as_dataframe(training_set.take(4), info)
```

For maximum control, manually display examples. Begin by taking some examples:

```
images, labels = [], []
for img, lbl in training_set.take(4):
  img = tf.squeeze(img)
  images.append(img), labels.append(lbl)
```

Continue by visualizing examples as shown in Listing 4-1.

Listing 4-1. Visualize Examples

```
import matplotlib.pyplot as plt

rows, cols = 2, 2
plt.figure(figsize=(10, 10))
for i in range(rows*cols):
  plt.subplot(rows, cols, i + 1)
  plt.imshow(images[i], cmap='bone')
```

```
  t = class_labels[labels[i]] + ' (' +\
      str(labels[i].numpy()) + ')'
  plt.title(t)
  plt.axis('off')
```

Now, we have an idea of what examples look like in the dataset.

Inspect Examples

Take some examples and convert them to NumPy arrays as shown in Listing 4-2.

Listing 4-2. Add NumPy Examples to Lists

```
features, labels = [], []
for img, lbl in training_set.take(4):
  img = tfds.as_numpy(img)
  lbl = tfds.as_numpy(lbl)
  features.append(img)
  labels.append(lbl)
```

It is easier to work with NumPy arrays when visualizing examples. Visualize as shown in Listing 4-3.

Listing 4-3. Visualize Inspected Examples

```
rows, cols = 2, 2
plt.figure(figsize=(10, 10))
for i in range(rows*cols):
  c = class_labels[labels[i]]
  s = str(features[i].shape)
  title = c + ' ' + s
  plt.subplot(rows, cols, i + 1)
  plt.title(title)
  plt.imshow(features[i], cmap='binary')
  plt.axis('off')
```

Reformat Images

Create a function to resize and scale images:

```
def format_image(image, label):
  image = tf.image.resize(image, (150, 150))/255.0
  return image, label
```

It is faster to train a model with smaller images. So resize images to 150 × 150 pixels. Scale images to improve training performance.

Note When resizing images to a smaller size, some information is lost. But the tf.image.resize does a good job of preserving as much information as possible. We resize images to 150 × 150 to preserve most of the information while increasing learning speed. Experiment with different sizes to see the impact on performance.

The function is used in the next section to facilitate building the input pipeline.

Build the Input Pipeline

The input pipeline is built by preparing the training, validation, and test sets for the learning model.

Set batch and shuffle parameters:

```
BATCH_SIZE = 200
SHUFFLE_SIZE = 500
```

Tip Experiment with batch and shuffle sizes to see how learning performance is impacted.

Transform data for optimum performance:

```
train_batches = training_set.shuffle(SHUFFLE_SIZE).\
map(format_image).batch(BATCH_SIZE).cache().prefetch(1)
```

```
validation_batches = validation_set.\
map(format_image).batch(BATCH_SIZE).cache().prefetch(1)

test_batches = test_set.\
map(format_image).batch(BATCH_SIZE).cache().prefetch(1)
```

By batching, training time is reduced. By shuffling, accuracy is usually increased. Caching helps better manage memory, and prefetching should reduce training time.

Inspect tensors:

```
train_batches, validation_batches, test_batches
```

Visualize and Inspect Examples from a Batch

Take the first training batch:

```
for img, lbl in train_batches.take(1):
  print (img.shape)
```

Inspect the first example from the batch:

```
img[0].shape, class_labels[lbl[0].numpy()]
```

The example contains a 150 × 150 × 3 image of either a “dog” or a “cat.” We can't know which because of randomization effects.

Extract four examples into lists:

```
images, labels = [], []
for i in range(4):
  tf.squeeze(img[i])
  images.append(img[i]), labels.append(lbl[i])
```

Squeeze out the *1* dimension because the *imshow()* function expects a 2D matrix.

Visualize images as shown in Listing 4-4.

Listing 4-4. Visualize Batched Examples

```
rows, cols = 2, 2
plt.figure(figsize=(10, 10))
for i in range(rows*cols):
  plt.subplot(rows, cols, i + 1)
```

```
  plt.imshow(images[i], cmap='bone')
  t = class_labels[labels[i]] + ' (' +\
      str(labels[i].numpy()) + ') ' +\
      str(images[i].shape)
  plt.title(t)
  plt.axis('off')
```

Build the Model

Get input shape:

```
for img, lbl in train_batches.take(1):
  in_shape = img.shape[1:]
in_shape
```

Import libraries:

```
from tensorflow.keras.models import Sequential
from tensorflow.keras.layers import Conv2D, MaxPooling2D,\
Dense, Flatten, Dropout
```

Clear and seed:

```
import numpy as np

tf.keras.backend.clear_session()
np.random.seed(0)
tf.random.set_seed(0)
```

Create a function to build the model as shown in Listing 4-5.

Listing 4-5. Multilayered CNN Model

```
def build_model():
  model = \
  Sequential([
  Conv2D(32, (3, 3), activation = 'relu',
         input_shape=in_shape, strides=1,
         kernel_regularizer='l1_l2'),
```

```
    MaxPooling2D(2, 2),
    Conv2D(64, (3, 3), activation='relu'),
    MaxPooling2D(2, 2),
    Conv2D(128, (3, 3), activation='relu'),
    MaxPooling2D(2),
    Conv2D(128, (3, 3), activation='relu'),
    MaxPooling2D(2, 2),
    Flatten(),
    Dense(512, activation='relu'),
    Dense(num_classes, activation='sigmoid')])
    return model
```

We created a function to hold the model for this experiment. We do this to show you how this can be done with a function.

The model begins with four Conv2D layers and four associated MaxPooling2D layers. A *Conv2D* layer computes a 2D convolution given input and 4D filter tensors. In our case, dimensions of tensors are (200, 150, 150, 3). Input tensors are batches of 200 150 × 150 color images. A *MaxPooling2D* layer performs a max pooling operation for 2D spatial data. So each max pooling operation reduces the dimensionality of the 2D spatial data it receives from each Conv2D layer. We use a pooling size of 2 × 2 because we experienced good performance with this size.

Convolutional layers apply filters to the original image or to other feature maps. The most important parameters are number and size of the kernels. *Pooling layers* perform max pooling or average pooling to reduce the dimensionality of the network. We use a pooling layer after each convolutional layer to reduce the dimensionality produced by the convolutional layer. Max pooling takes the maximum value in a filter region, while average pooling takes the average value in a filter region.

The first Conv2D layer accepts the number of kernels, kernel size, input shape, strides, and regularization. We use a kernel size of 3 × 3 because we experienced good performance with this size. We added l1 and l2 regularization to reduce overfitting. A *kernel* is a filter used to extract features from images. Specifically, a kernel is a matrix that moves over the input data, performs the dot product with the subregion of input data, and gets the output as the matrix of dot products (or feature maps). A *feature map* captures the result of applying the filters (or kernels) to an input image.

The reason for visualizing a feature map for a specific input image is to try to gain some understanding of what features our CNN detects. We apply several convolutional layers to the data because we hope that each layer produces feature maps that help us detect clearer and clearer features.

The remaining Conv2D layers increase the kernel size to improve model performance. After each Conv2D layer is a MaxPooling2D to reduce the dimensionality contributed by convolution. Each Conv2D layer uses Rectified Linear Unit (ReLU) activation, which combats the vanishing gradient problem occurring with sigmoid distributions.

A Flatten layer prepares output from the convolutional layers for the Dense layers. Dense layers require a single long-feature vector as input. Two fully connected Dense layers are included for classification of the output.

Create the model:

```
cat_dog_model = build_model()
```

Inspect the model:

```
cat_dog_model.summary()
```

For a great resource on applying regularization to a learning model, peruse *www.machinecurve.com/index.php/2020/01/23/how-to-use-l1-l2-and-elastic-net-regularization-with-keras/*

Compile and Train the Model

Compile:

```
loss = tf.keras.losses.SparseCategoricalCrossentropy(
    from_logits=True)

cat_dog_model.compile(optimizer='adam',
              loss=loss,
              metrics=['accuracy'])
```

Since the model outputs logits, set *from_logits=True*. The model outputs logits because we used the sigmoid activation function in the output layer of the model.

Train:

```
epochs = 10
history = cat_dog_model.fit(
    train_batches, epochs=epochs, verbose=1,
    validation_data=validation_batches)
```

The model learns from the training set and validates with the validation set. The model never touches the test set.

Evaluate the Model for Generalizability

Use the evaluate method with the test batch:

```
metrics = cat_dog_model.evaluate(test_batches)
```

The *evaluate()* method returns the loss and accuracy values. We use *test_batches* to evaluate for generalizability because the model has never seen this dataset. **Generalization** is a learning model's ability to adapt properly to new, previously unseen data, drawn from the same distribution as the one used to create the model.

Visualize Performance

Create a visualization function as shown in Listing 4-6.

Listing 4-6. Plot Model Performance

```
def viz(hd):
  acc = hd['accuracy']
  val_acc = hd['val_accuracy']
  loss = hd['loss']
  val_loss = hd['val_loss']
  plt.figure(figsize=(8, 8))
  plt.subplot(1, 2, 1)
  plt.plot(acc, label='Training Accuracy')
  plt.plot(val_acc, label='Validation Accuracy')
  plt.legend(loc='lower right')
  plt.title('Training and Validation Accuracy')
```

```
  plt.subplot(1, 2, 2)
  plt.plot(loss, label='Training Loss')
  plt.plot(val_loss, label='Validation Loss')
  plt.legend(loc='upper right')
  plt.title('Training and Validation Loss')
  plt.show()
```

Invoke the function:

```
viz(history.history)
```

The model is overfitting because the test accuracy is not that closely aligned with the training accuracy.

Augmentation with Preprocessing Layers

To mitigate overfitting and improve model performance, we experiment with data augmentation. Apply random horizontal flips, random rotation, and random zoom as shown in Listing 4-7.

Listing 4-7. Keras Preprocessing Layers for Transformation

```
from tensorflow.keras import layers

data_augmentation = tf.keras.Sequential(
  [
    layers.experimental.preprocessing.RandomFlip('horizontal'),
    layers.experimental.preprocessing.RandomRotation(0.1),
    layers.experimental.preprocessing.RandomZoom(0.1),
  ]
)
```

We chose to use these layers based on trial and error experimentation. Feel free to experiment with other layers.

Note This transformation operation is experimental, which means that it may change in the future.

Display an augmented image as shown in Listing 4-8.

Listing 4-8. Display an Augmented Image

```
plt.figure(figsize=(10, 10))
for images, _ in train_batches.take(1):
  for i in range(9):
    augmented_images = data_augmentation(images)
    ax = plt.subplot(3, 3, i + 1)
    plt.imshow(augmented_images[0])
    plt.axis('off')
```

The visualization shows data augmentation applied to the same image several times. Change the index from 0 (in *augmented_images[0]*) to between 1 and 199 (batch size 200) to see different images in the *plt.imshow* code statement.

Build the Model

Build the model by clearing previous sessions, generating a seed, and creating the model for the experiment.

Clear and seed:

```
tf.keras.backend.clear_session()
np.random.seed(0)
tf.random.set_seed(0)
```

Create the model as shown in Listing 4-9.

Listing 4-9. Model with Augmentation Layer

```
cat_dog_layers = Sequential([
  data_augmentation,
  Conv2D(32, (3, 3), activation = 'relu',
         input_shape=in_shape, strides=1,
         kernel_regularizer='l1_l2'),
  MaxPooling2D(2, 2),
  Conv2D(64, (3, 3), activation='relu'),
  MaxPooling2D(2, 2),
```

```
    Conv2D(128, (3, 3), activation='relu'),
    MaxPooling2D(2),
    Conv2D(128, (3, 3), activation='relu'),
    MaxPooling2D(2, 2),
    Flatten(),
    Dense(512, activation='relu'),
    Dense(num_classes, activation='sigmoid')
])
```

Notice that the first layer is the augmentation that we created! So the model gets the augmentations from the first layer and continues like other convolutional models from that point forward.

Compile and Train the Model

Compile:

```
cat_dog_layers.compile(
    optimizer='adam',
    loss=tf.keras.losses.SparseCategoricalCrossentropy(
        from_logits=True),
    metrics=['accuracy'])
```

Train:

```
epochs = 10
history = cat_dog_layers.fit(
    train_batches, epochs=epochs, verbose=1,
    validation_data=validation_batches)
```

Evaluate the Model for Generalizability

Evaluate:

```
metrics = cat_dog_layers.evaluate(test_batches)
```

Data augmentation doesn't guarantee better performance. It is only a technique that can be applied for that purpose. In addition, we run this model for only ten epochs. So it may take more epochs to see better performance.

Visualize Performance

Visualize training performance:

```
viz(history.history)
```

Overfitting has been somewhat mitigated. Try running the experiment for more epochs to see what happens.

Apply Data Augmentation on Images

Apply data augmentation by performing transformations directly on images.

Create functions to augment images as shown in Listing 4-10.

Listing 4-10. Functions That Apply Augmentations on Images

```
def random_crop(image):
    shape = tf.shape(image)
    min_dim = tf.reduce_min([shape[0], shape[1]]) * 90 // 100
    return tf.image.random_crop(image, [min_dim, min_dim, 3])

def preprocess(image, label):
  cropped_image = random_crop(image)
  cropped_image = tf.image.random_flip_left_right(cropped_image)
  resized_image = tf.image.resize(cropped_image, [150, 150])
  final_image = tf.keras.applications.xception.preprocess_input(
      resized_image)
  return final_image, label
```

The *random_crop* function randomly crops images. The *preprocess* function calls random_crop, randomly flips images left to right, and resizes them. The *tf.keras.applications.xception.preprocess_input* utility preprocesses a tensor (or NumPy array) by encoding a batch of images. Through research and experimentation, we found that these augmentations increase performance and mitigate overfitting. Experiment with different augmentations and see what happens.

Build the Input Pipeline

Import the *partial* package:

```
from functools import partial
```

The partial package allows us to create partial functions. Partial functions allow us to fix a certain number of arguments of a function and generate a new function.

Set batch and shuffle size parameters:

```
BATCH_SIZE = 200
SHUFFLE_SIZE = 500
```

Tip Experiment with batch and shuffle sizes to see how learning performance is impacted.

Build the pipeline:

```
train_shuffle = training_set.shuffle(1000)
train_batches = train_shuffle.map(partial(preprocess)).\
  batch(BATCH_SIZE).prefetch(1)
validation_batches = validation_set.map(preprocess).\
  batch(BATCH_SIZE).prefetch(1)
test_batches = test_set.map(preprocess).\
  batch(BATCH_SIZE).prefetch(1)
```

Setting partial on the preprocess function fixes its arguments so we can apply transformations to images *only from the training set*! We don't need to augment the other datasets because the model only learns from the training set.

Display an Augmented Image

Here's what the augmentations do to an image as shown in Listing 4-11.

Listing 4-11. Visualize Augmentations on an Image

```
plt.figure(figsize=(10, 10))
for images, _ in train_batches.take(1):
  for i in range(9):
    augmented_images = data_augmentation(images)
    Images = np.clip(augmented_images, 0, 1)
    ax = plt.subplot(3, 3, i + 1)
    plt.imshow(Images[0])
    plt.axis('off')
```

We display augmentations on the first image from the batch. Experiment with the index number to see augmentations of any of the 200 images from the batch. Be sure to choose an index between 0 and 199 because the batch size is 200!

Build the Model

Clear and seed:

```
tf.keras.backend.clear_session()
np.random.seed(0)
tf.random.set_seed(0)
```

Create the model:

```
cat_dog_images = build_model()
```

Compile and Train the Model

Compile:

```
cat_dog_images.compile(
    optimizer='adam',
    loss=tf.keras.losses.SparseCategoricalCrossentropy(
        from_logits=True),
    metrics=['accuracy'])
```

Train:

```
epochs = 5
history = cat_dog_images.fit(
    train_batches, epochs=epochs,
    verbose=1, validation_data=validation_batches)
```

We only train for a few epochs because training time is substantially increased with the augmentations that we applied to images. Experiment with the number of epochs to see what happens.

Evaluate the Model for Generalizability

Evaluate:

```
cat_dog_images.evaluate(test_batches)
```

We use the test set because it has never been seen by the model.

Note The accuracy may seem to be less, but it actually isn't. The reason is that we may need to run the model for many more epochs. This is why we encourage you to experiment with the number of epochs.

Visualize Performance

Visualize training performance:

```
viz(history.history)
```

Validation accuracy trajectory is much better! That is, the training and test values are more closely aligned.

Predictions

To complete the experiment, let's see how well the model predicts from the test set. We can make predictions on the entire test dataset or on a batch from the dataset.

Begin by making predictions on the entire test dataset and displaying the prediction array for the first example:

```
predictions =cat_dog_images.predict(test_batches)
list(predictions[0])
```

Predict on *test_batches* because it has never been seen by the model. The *predict()* method returns NumPy arrays of predictions. For classification experiments, each element in a prediction array represents a class label. Since we have two classes in our experiment (cats and dogs), each prediction array contains two elements. The cat prediction is the first value and the dog prediction is the second value in the array. Each value is the probability of the prediction. The magnitude of the probability represents the likelihood that the element is the predicted class. So the higher probability between the two elements in the prediction array represents the prediction. For example, a prediction array of [0.33, 0.778] means that the prediction is a dog since it has the higher probability.

Return the prediction for the first example:

```
first_prediction = tf.math.argmax(predictions[0])
class_labels[first_prediction.numpy()]
```

The *tf.math.argmax()* API returns the index with the largest value across axes of a tensor. Simply, it returns the index in the prediction array with the highest value.

We can also make predictions on a single batch:

```
first_batch = test_batches.take(1)
predict_batch = cat_dog_images.predict(first_batch)
```

Take the first batch and make predictions on it.

Get the prediction for the first example in the batch:

```
first_batch_prediction = tf.math.argmax(predict_batch[0])
class_labels[first_batch_prediction.numpy()]
```

Get the first actual label:

```
for image, label in first_batch:
  print ('batch size:', image.shape[0])
class_labels[label[0].numpy()]
```

Grab actual images and labels from the first batch. Display the first actual image shape and its corresponding label. If the prediction matches the actual label, the prediction is correct!

Manually check prediction accuracy as shown in Listing 4-12.

Listing 4-12. Check Prediction Accuracy

```
cnt = 0
for i in range(image.shape[0]):
  pred = tf.math.argmax(predict_batch[i]).numpy()
  actual = label[i].numpy()
  if actual == pred:
    cnt += 1
acc = cnt / image.shape[0]
'{percent:.2%}'.format(percent=acc) + ' accuracy'
```

The code computes the prediction accuracy of the test data and displays it as a percentage.

Visualize Predictions

Grab the first batch of images and labels from the test set:

```
for img, lbl in test_batches.take(1):
  print (img.shape)
```

Each batch contains 200 150 × 150 color images.

Inspect the first image from the first batch:

```
img[0].shape, class_labels[lbl[0].numpy()]
```

Visualize the image:

```
Image= np.clip(img[0], 0, 1)
fig = plt.imshow(Image)
fig = plt.axis('off')
```

We use the *tf.keras.applications.xception.preprocess_input* utility to preprocess images (see the preprocessing function in Listing 4-10). The utility instantiates the Xception architecture for optimum performance. However, it doesn't scale images. So we must transform image pixels to between 0 and 1 to display them appropriately.

Process some examples:

```
images, labels = [], []
for i in range(20):
  tf.squeeze(img[i])
  images.append(img[i]), labels.append(lbl[i])
```

Create a function to display a set of images and labels. The function determines if a prediction is correct or incorrect. Titles for correct predictions are displayed normally, but titles for incorrect predictions are displayed in red text.

Create the function as shown in Listing 4-13.

Listing 4-13. Function to Display Predictions Against Actual Labels

```
def display_test(feature, target, num_images,
                 n_rows, n_cols, cl, p):
  for i in range(num_images):
    plt.subplot(n_rows, 2*n_cols, 2*i+1)
    Image= np.clip(feature[i], 0, 1)
    plt.imshow(Image)
    pred = cl[tf.math.argmax(p[i]).numpy()]
    actual = cl[target[i]]
    title_obj = plt.title(actual + ' (' +\
                          pred + ') ')
    if pred == actual:
      title_obj
    else:
      plt.getp(title_obj, 'text')
      plt.setp(title_obj, color='r')
    plt.tight_layout()
    plt.axis('off')
```

Invoke the function:

```
num_rows, num_cols = 5, 4
num_images = num_rows*num_cols
plt.figure(figsize=(20, 20))
display_test(images, labels, num_images, num_rows,
             num_cols, class_labels, predictions)
```

Images are displayed with titles. Each title shows the actual label with the prediction in parentheses. If the prediction was incorrect, the title is shown in red.

Note Although the learning models we apply may appear complex to the novice deep learning aficionado, they are really very simple. As you gain experience with learning models, you will gain the confidence to build more complex ones or migrate to transfer learning models. We cover transfer learning models in this book.

An Experiment with rock_paper_scissors

Although we just worked through an experiment with a large and complex dataset, we believe that practicing deep learning with multiple datasets is critically important to becoming a competent practitioner. During our seminars, we find that participants learn much faster when they can apply what they have learned to another context. In this case, it is a new dataset. The experiment is conducted on the rock_paper_scissors dataset, which contains 2,892 images of hands playing the rock-paper-scissors game.

Configure TensorBoard

TensorBoard is a tool that provides measurements and visualizations during the machine learning workflow. It tracks metrics such as loss, accuracy, model graph visualization, and embedding projections to a lower-dimensional space. We expose you to TensorBoard for this experiment to provide access to another visualization tool.

For a nice tutorial, peruse

www.tensorflow.org/tensorboard/get_started

Load the TensorBoard notebook extension:

```
%load_ext tensorboard
```

Import the requisite library that we use in the experiment:

```
import datetime
```

Clear logs from previous runs:

```
!rm -rf ./logs/
```

Load Data

Load the train and test sets:

```
(train_digits, test_digits), rps_info = tfds.load(
    'rock_paper_scissors', with_info=True,
    data_dir='tmp', as_supervised=True,
    split=['train', 'test'])
```

Inspect the Data

We believe it is a good idea to explore a new dataset to get an idea of its configuration. So we get the metadata object, visualize some examples, and get the number of training and test examples, example shapes, and label names. We need to know all of this information to conduct the experiment.

Get metadata:

```
rps_info
```

Visualize:

```
fig = tfds.show_examples(train_digits, rps_info)
```

Get examples and number of labels:

```
train_examples = rps_info.splits['train'].num_examples
test_examples = rps_info.splits['test'].num_examples
num_labels = rps_info.features['label'].num_classes
train_examples, test_examples, num_labels
```

Get shapes:

```
rps_info.features['image'].shape,\
rps_info.features['label'].shape
```

Get label names:

```
label_name = rps_info.features['label'].int2str
for lbl in range(num_labels):
  print (label_name(lbl), end=' ')
```

Inspect image shape:

```
for image, label in train_digits.take(5):
  print (image.shape)
```

Preprocess the Data

The next step is to preprocess the data for consumption by the learning model. We begin by getting a value for half the size of the current image. We use this later to reshape images. We chose this size to reduce training time. By choosing this size, we found that the model still performed well. We encourage you to experiment with different image sizes, to see how learning is impacted. We also scale and resize images. Remember that learning models work better with smaller pixel sizes. Scaling is the best way to reduce image size without adversely impacting learning.

Create a shape that halves image size:

```
new_pixels = rps_info.features['image'].shape[0] // 2
new_pixels
```

Create a preprocessing function:

```
def format_digits(image, label):
  image = tf.cast(image, tf.float32) / 255.
  image = tf.image.resize(image, [new_pixels, new_pixels])
  return image, label
```

Preprocess train and test sets:

```
train_original = train_digits.map(format_digits)
test_original = test_digits.map(format_digits)
```

Explore an example:

```
for image, label in train_original.take(1):
  finger_img_shape = image.shape
  print (image.shape, image[0][0].numpy(), label.numpy())
```

We always explore at least one example after it is preprocessed to see how it looks.

Visualize Processed Data

Create a visualization function as shown in Listing 4-14.

Listing 4-14. Visualization Function

```
def preview_dataset(dataset):
  plt.figure(figsize = (12, 12))
  plot_index = 0
  for image, label in dataset.take(12):
    plot_index += 1
    plt.subplot(3, 4, plot_index)
    plt.axis('Off')
    label = label_name(label.numpy())
    plt.title(label)
    plt.imshow(image.numpy())
```

Invoke the function:

```
preview_dataset(train_original)
```

Augment Training Data

We include data augmentation to improve learning. We chose the augmentations based on experimentation. We separate different augmentations into separate functions for flexibility. We conduct the experiment with all of the augmentations, but you can use one or more of the functions to conduct a new experiment. You can even skip the augmentation step to see how well the model learns with un-augmented training data.

Augmentation is an art. We haven't found any applied deep learning research on how to systematically augment a given dataset. Deep learning algorithms have been around for many years, but application of them is much more recent. A big reason is that computing power was not even close to what it is now. And it is much cheaper! Another reason is that TensorFlow has only been available as an open source product for a few years.

Create Data Augmentation Functions

We now create functions to flip, color, rotate, invert, and zoom images.

Create a function to randomly flip an image:

```
def flip(image: tf.Tensor) -> tf.Tensor:
  image = tf.image.random_flip_left_right(image)
  image = tf.image.random_flip_up_down(image)
  return image
```

Note Convert images to tensors with the *tf.Tensor* API.

Create a function to randomly augment color as shown in Listing 4-15.

Listing 4-15. Randomly Augment Image Color

```
def paint(image: tf.Tensor) -> tf.Tensor:
  image = tf.image.random_hue(image, max_delta=0.2)
  image = tf.image.random_saturation(image, lower=0.7, upper=1.3)
  image = tf.image.random_brightness(image, 0.05)
  image = tf.image.random_contrast(image, lower=0.8, upper=1)
  image = tf.clip_by_value(image, clip_value_min=0,
                           clip_value_max=1)
  return image
```

Create a function to randomly rotate an image:

```
def rotate(image: tf.Tensor) -> tf.Tensor:
  return tf.image.rot90(
      image,
```

```
    tf.random.uniform(
        shape=[], minval=0,
        maxval=4, dtype=tf.int32))
```

Create a function to randomly invert an image:

```
def invert(image: tf.Tensor) -> tf.Tensor:
  random = tf.random.uniform(shape=[], minval=0, maxval=1)
  if random > 0.5:
    image = tf.math.multiply(image, -1)
    image = tf.math.add(image, 1)
  return image
```

Create a function to zoom an image as shown in Listing 4-16.

Listing 4-16. Zoom Augmentation

```
def zoom(
    image: tf.Tensor, min_zoom=0.8, max_zoom=1.0) -> tf.Tensor:
  image_width, image_height, image_colors = image.shape
  crop_size = (image_width, image_height)
  scales = list(np.arange(min_zoom, max_zoom, 0.01))
  boxes = np.zeros((len(scales), 4))
  for i, scale in enumerate(scales):
    x1 = y1 = 0.5 - (0.5 * scale)
    x2 = y2 = 0.5 + (0.5 * scale)
    boxes[i] = [x1, y1, x2, y2]
  def random_crop(img):
    crops = tf.image.crop_and_resize(
        [img], boxes=boxes,
        box_indices=np.zeros(len(scales)),
        crop_size=crop_size)
    return crops[tf.random.uniform(
        shape=[], minval=0, maxval=len(scales), dtype=tf.int32)]
  choice = tf.random.uniform(
      shape=[], minval=0., maxval=1., dtype=tf.float32)
  return tf.cond(
      choice < 0.5, lambda: image, lambda: random_crop(image))
```

The zoom function generates crop settings ranging from 1% to 20%. It then creates bounding boxes to hold the cropped images. Returned cropped images are resized to keep the size uniform for training. Cropping is only performed 50% of the time.

Create an augment function:

```
def augment_data(image, label):
  image = flip(image)
  image = paint(image)
  image = rotate(image)
  image = invert(image)
  image = zoom(image)
  return image, label
```

To experiment, remove one or more of the augmentations by commenting out the function.

Augment Train Data

Map augmentations to train data:

```
train_augmented = train_original.map(augment_data)
```

Visualize training example augmentations:

```
preview_dataset(train_augmented)
```

Visualize the original test set:

```
preview_dataset(test_original)
```

Build the Input Pipeline

Finish the pipeline:

```
BATCH_SIZE = 32

train_fingers = train_augmented.shuffle(train_examples).cache().\
  batch(BATCH_SIZE).prefetch(tf.data.experimental.AUTOTUNE)

test_fingers = test_original.batch(BATCH_SIZE)
```

The prefetch API enables the input pipeline to asynchronously fetch batches while the model is training.

Create the Model

Clear and seed:

```
tf.keras.backend.clear_session()
np.random.seed(0)
tf.random.set_seed(0)
```

Verify image shape:

```
finger_img_shape
```

Build the model as shown in Listing 4-17.

Listing 4-17. Create the Model

```
finger_model = Sequential([
  Conv2D(64, 3, activation='relu',
         input_shape=finger_img_shape,
         kernel_regularizer='l1_l2'),
  MaxPooling2D(2, 2),
  Conv2D(64, 3, activation='relu'),
  MaxPooling2D(2, 2),
  Conv2D(128, 3, activation='relu'),
  MaxPooling2D(2, 2),
  Conv2D(128, 3, activation='relu'),
  MaxPooling2D(2, 2),
  Flatten(),
  Dense(512, activation='relu'),
  Dense(num_labels, activation='softmax')])
```

Inspect the model:

```
tf.keras.utils.plot_model(
    finger_model,
    show_shapes=True,
    show_layer_names=True)
```

We introduce another utility to inspect the model as an alternative. We try to expose you to as many options as possible. But we don't want to overwhelm. So we sprinkle in something new within the context of an experiment. We have found that learning a complex software like TensorFlow is much faster done within some context.

Compile and Train

Compile:

```
optimizer = tf.keras.optimizers.RMSprop(learning_rate=0.001)

finger_model.compile(
    optimizer=optimizer,
    loss=tf.keras.losses.sparse_categorical_crossentropy,
    metrics=['accuracy'])
```

Establish training parameters:

```
steps_per_epoch = train_examples // BATCH_SIZE
validation_steps = test_examples // BATCH_SIZE
print('steps_per_epoch:', steps_per_epoch)
print('validation_steps:', validation_steps)
```

Remove logs and checkpoints to clear out previous model sessions:

```
!rm -rf tmp/checkpoints
!rm -rf logs
```

Prepare a TensorBoard callback:

```
log_dir = 'logs/fit/' +\
  datetime.datetime.now().strftime('%Y%m%d-%H%M%S')
tensorboard_callback = tf.keras.callbacks.TensorBoard(
    log_dir=log_dir, histogram_freq=1)
```

Train:

```
training_history = finger_model.fit(
    train_fingers.repeat(),
    validation_data=test_fingers.repeat(),
    epochs=10,
```

```
    steps_per_epoch=steps_per_epoch,
    validation_steps=validation_steps,
    callbacks=[tensorboard_callback])
```

Visualize Performance

Create a function as shown in Listing 4-18.

Listing 4-18. Visualization Function

```
def viz_history(training_history):
  loss = training_history.history['loss']
  val_loss = training_history.history['val_loss']
  accuracy = training_history.history['accuracy']
  val_accuracy = training_history.history['val_accuracy']
  plt.figure(figsize=(14, 4))
  plt.subplot(1, 2, 1)
  plt.title('Loss')
  plt.xlabel('Epoch')
  plt.ylabel('Loss')
  plt.plot(loss, label='Training set')
  plt.plot(val_loss, label='Test set', linestyle='--')
  plt.legend()
  plt.grid(linestyle='--', linewidth=1, alpha=0.5)
  plt.subplot(1, 2, 2)
  plt.title('Accuracy')
  plt.xlabel('Epoch')
  plt.ylabel('Accuracy')
  plt.plot(accuracy, label='Training set')
  plt.plot(val_accuracy, label='Test set', linestyle='--')
  plt.legend()
  plt.grid(linestyle='--', linewidth=1, alpha=0.5)
  plt.show()
```

Invoke the function:

```
viz_history(training_history)
```

Launch TensorBoard:

```
%tensorboard --logdir logs/fit
```

Again, we show you how to use TensorBoard within the context of an experiment to speed your learning. We don't use TensorBoard that often because we can see model performance with just a few lines of code as we have done previously. But we believe it is a good idea to expose you to the product. You may prefer it over the way we visualize learning performance.

Close TensorBoard Server

Use the Global Regular Expression Print (grep) command to get the running TensorBoard process details:

```
!ps -ef | grep tensorboard
```

Output looks something like this:

```
root        10757    4202 26 19:13 ?        00:00:04 python3 /usr/local/bin/
tensorboard --logdir logs/fit
root        10794    4202  0 19:13 ?        00:00:00 /bin/bash -c ps -ef |
grep tensorboard
root        10796   10794  0 19:13 ?        00:00:00 grep tensorboard
```

The first number 10757 is the current process identifier (pid) for TensorFlow.

Kill the process using the pid:

```
!kill 10757
```

Note The pid changes each time TensorBoard is activated.

CHAPTER 5

Introduction to Tensor Processing Units

We introduce you to Tensor Processing Units with code examples. A **Tensor Processing Unit (TPU)** is an application-specific integrated circuit (ASIC) designed to accelerate ML workloads. The TPUs available in TensorFlow are custom-developed from the ground up by the Google Brain team based on its plethora of experience and leadership in the ML community. *Google Brain* is a deep learning artificial intelligence (AI) research team at Google who research ways to make machines intelligent for the improvement of people's lives.

Notebooks for chapters are located at the following URL:

https://github.com/paperd/deep-learning-models

TPU Hardware Accelerator

We demonstrate learning experiments with the TPU hardware accelerator available from the Google Colab cloud service. Our examples do not demonstrate the full efficacy of a TPU, but are provided to get you started. We use high-level TensorFlow APIs to get models running on Cloud TPU hardware.

The hardware accelerator we use in Colab is commonly referred to as Cloud TPU. *Cloud TPU* is designed to run cutting-edge ML models with AI services on Google Cloud. It enables processing of ML workloads on Google's TPU accelerator hardware using TensorFlow. Its custom high-speed network offers over 100 petaflops of performance in a single pod, which is enough computational power to transform a business to AI-ready or create the next research breakthrough. Cloud TPU is designed for maximum performance and flexibility to help researchers, developers, and businesses build TensorFlow compute clusters that can leverage CPUs, GPUs, and TPUs.

For a comprehensive guide to deploying Cloud TPU, peruse

D. Paper, *State-of-the-Art Deep Learning Models in TensorFlow*, https://doi.org/10.1007/978-1-4842-7341-8_5

https://cloud.google.com/tpu/docs/tpus

For all available tutorials on Cloud TPU (as of this writing), peruse *https://cloud.google.com/tpu/docs/tutorials*

Advantages of Cloud TPU

Cloud TPU resources accelerate the performance of linear algebra computation, which is used heavily in ML applications. Cloud TPU minimizes the time-to-accuracy when we train large, complex neural network models. Models that previously took weeks to train on other hardware platforms can converge in hours on Cloud TPU.

When to Use Cloud TPU

Cloud TPU is optimized for specific workloads. In some situations, we use GPUs or CPUs to run ML workloads. In general, decide what hardware to use based on the following guidelines.

CPUs

* Quick prototyping that requires maximum flexibility
* Simple models that do not take long to train
* Small models with small effective batch sizes
* Models dominated by custom TensorFlow operations written in C++
* Models limited by available I/O or the networking bandwidth of the host system

GPUs

* Models where source code doesn't exist or is too onerous to change
* Models with a significant number of custom TensorFlow operations that must run at least partially on CPUs
* Models with TensorFlow ops that are not available on Cloud TPU
* Medium-to-large models with larger effective batch sizes

TPUs

* Models dominated by matrix computations

* Models with no custom TensorFlow operations inside the main training loop

* Models that train for weeks or months

* Very large models with very large effective batch sizes

Cloud TPU is not suited for linear algebra programs that require frequent branching or are dominated element-wise by algebra. Cloud TPU is optimized to perform fast on bulky matrix multiplication. So a workload that is not dominated by matrix multiplication is unlikely to perform well on Cloud TPU compared to other platforms. Workloads that require high-precision arithmetic are not suitable for Cloud TPU (e.g., double-precision arithmetic). Workloads that access memory in a sparse manner might not be available on Cloud TPU.

For a great resource on Cloud TPU, peruse *https://cloud.google.com/tpu/docs/tpus*

Import the TensorFlow Library

Import the library and alias it as **tf**:

```
import tensorflow as tf
```

Enable TPU Runtime

It's very easy to enable the TPU in a Colab notebook:

1. Click *Runtime* in the top-left menu.
2. Click *Change runtime type* from the drop-down menu.
3. Choose *TPU* from the *Hardware accelerator* drop-down menu.
4. Click *Save*.

Note The TPU must be enabled in *each* notebook. But it only has to be enabled once.

TPU Detection

Set up the TPU resolver and verify that it is running:

```
tpu = tf.distribute.cluster_resolver.TPUClusterResolver()
print('Running on TPU ', tpu.cluster_spec().as_dict()['worker'])
```

The message looks something like this:

RUNNING ON TPU ['10.112.96.162:8470']

Tip If you get the error **NAME 'TF' IS NOT DEFINED**, re-execute the code to import the TensorFlow library! For some reason, we sometimes have to re-execute the TensorFlow library import in Colab. We don't know why this is the case.

The *tf.distribute.cluster_resolver.TPUClusterResolver()* API is the Cluster Resolver for Google Cloud TPU. *Cluster Resolvers* provide a way for TensorFlow to communicate with various cluster management systems (e.g., GCE, AWS, etc.) and access necessary information to set up distributed training. By letting TensorFlow communicate with these systems, it is able to automatically discover and resolve IP addresses for various TensorFlow workers. So it is able to eventually automatically recover from underlying machine failures and scale TensorFlow worker clusters up and down system pipelines.

Configure the TPU for This Notebook

Make devices on the cluster available to use:

```
tf.config.experimental_connect_to_cluster(tpu)
```

Initialize TPU devices:

```
tf.tpu.experimental.initialize_tpu_system(tpu)
```

Note Both configurations are experimental, which means that they may be changed in the future.

Create a Distribution Strategy

A **distribution strategy** is an abstraction used to distribute training across multiple CPUs, GPUs, or TPUs. To change how a model runs on a given device, simply swap out the distribution strategy. The *tf.distribute.Strategy* is the TensorFlow API for applying a distribution strategy to a model. Applying this API allows us to distribute training for existing models across multiple devices with minimal code changes.

The API is designed with three key goals in mind:

* Easy to use and support for multiple user segments (e.g., data scientists)

* Provides good performance out of the box

* Easy switching between strategies

Create a TPU strategy for this notebook:

```
tpu_strategy = tf.distribute.TPUStrategy(tpu)
```

A list of available TPUs and other devices is displayed.

Manual Device Placement

After the TPU system is initialized, we can use manual device placement to direct computation on a single TPU device as shown in Listing 5-1.

Listing 5-1. Run Computations on TPU Devices

```
a = [[1.0, 2.0, 3.0], [4.0, 5.0, 6.0]]
b = [[7.0, 8.0], [9.0, 10.0], [11.0, 12.0]]
with tf.device('/TPU:7'):
  c = tf.matmul(a, b)
I = [[1.0, 0.0], [0.0, 1.0]]
```

```
with tf.device('/TPU:6'):
  d = tf.matmul(c, I)
print('c device:', c.device)
print(c)
print('d device:', d.device)
print(d)
```

Multiply matrix a by matrix b and place the result in matrix c within the scope of TPU device 7. Next, multiply matrix c by the Identity matrix and place the result in matrix d within the scope of TPU device 6.

Run a Computation in All TPU Cores

To replicate a computation so it can run in all TPU cores, pass it to the *strategy.run* API:

```
@tf.function
def matmul_fn(x, y):
  z = tf.matmul(x, y)
  return z

z = tpu_strategy.run(matmul_fn, args=(a, b))
print(z)
```

Create a function that multiplies two matrices. If eager behavior is enabled, make the function a *tf.function* or call *strategy.run* inside a tf.function. Eager behavior is automatically enabled in Colab!

Invoke the function with the *strategy.run* API we create, which ensures that all the TPU cores obtain the same inputs (a, b) and matmul is applied on each core independently. The outputs are the values from all of the core replicas.

Eager Execution

Eager execution must be enabled for Cloud TPU to operate. *Eager execution* (or eager behavior) is an imperative programming environment because it evaluates operations immediately without building graphs. So TensorFlow operations return concrete values instead of constructing a computational graph to run later. It also offers an intuitive interface that allows you to structure code naturally and use Python data structures. Other benefits include quick iteration on small models and small data and easier debugging because you can call ops directly to inspect running models and test changes.

Note Eager execution is enabled by default in TensorFlow 2.x.

Experiments

We include four Cloud TPU experiments. Experiments begin with simple datasets and progress to more complex ones. The importance of working with multiple datasets is to gain experience and confidence. Each dataset is different, so the experiments are a bit different. The learning process is generally the same, but each dataset has different characteristics. The main reason we present multiple experiences is to give you more practice. We have found, through working with seminar participants, that they learn faster by working with multiple datasets and they also enjoy the learning experience much more!

Digits Experiment

The *digits* dataset is embedded in the *sklearn.datasets* package. It consists of 1,797 8 × 8 images of handwritten digits from 0 to 9. *Scikit-learn* (also known as sklearn) is a ML library in Python. It contains a few small standard datasets (e.g., digits) that do not require to download any file from an external website.

Import requisite libraries:

```
from sklearn.datasets import load_digits
from sklearn.model_selection import train_test_split
import matplotlib.pyplot as plt
```

Load the dataset:

```
digits = load_digits()
```

Get the keys:

```
digits.keys()
```

Keys include "data," "target," "target_names," "images," and "DESCR." Key *data* contains images as a flattened data matrix of shape (1797, 64). Key *target* contains labels as a vector of shape (1797,). Key *target_names* is a list of the names of target classes. Key *images* contains images as a matrix of shape (1797, 8, 8). Key *DESCR* contains the full description of the dataset. So digits contains 1,797 images of 8 × 8 pixels and 1,797 corresponding scalar labels.

The reason that this dataset has keys is because it is a scikit-learn (sklearn) dataset. It features various classification, regression, and clustering algorithms including support vector machines, random forests, gradient boosting, k-means, and DBSCAN. It is designed to interoperate with the Python numerical and scientific libraries NumPy and SciPy. It also contains practice datasets like digits.

Display an image:

```
images = digits.images
image = images[0]
fig = plt.imshow(image, cmap='binary')
fig = plt.axis('off')
```

Preprocess the Data

Since digits is a scikit-learn practice set, we can preprocess it very easily. The *train_test_split* method automatically splits data into training and test sets. You can adjust the split ratio with the *test_size* parameter.

Create a function to preprocess the data as shown in Listing 5-2.

Listing 5-2. Function to Preprocess Data

```
def load_data(digits, splits, random, scale):
  X = digits.images
  y = digits.target
  x_train, x_test, y_train, y_test = train_test_split(
    X, y, test_size=splits, random_state=random)
  x_train, x_test = x_train / scale, x_test / scale
  return (x_train, y_train), (x_test, y_test)
```

Preprocess the data:

```
splits, seed, scale = 0.33, 0, 255.0

(x_train, y_train), (x_test, y_test) = load_data(
    digits, splits, seed, scale)
```

Data split is 67% training and 33% test. Of course, you can adjust it to be a different split by changing the 0.33 to another value.

Get target names and number of classes:

```
target_names = digits.target_names
num_classes = len(target_names)
num_classes, target_names
```

We take advantage of the built-in sklearn method *target_names* to get the target (class label) names.

Build the Input Pipeline

Prepare data for TensorFlow consumption:

```
train_dataset = tf.data.Dataset.from_tensor_slices(
    (x_train, y_train))
test_dataset = tf.data.Dataset.from_tensor_slices(
    (x_test, y_test))
```

Inspect training tensors:

```
for img, lbl in train_dataset.take(1):
  print (img.shape, lbl)
```

Set parameters and build the pipeline:

```
BATCH_SIZE = 64
SHUFFLE_BUFFER_SIZE = 100

train_ds = train_dataset\
 .shuffle(SHUFFLE_BUFFER_SIZE)\
 .batch(BATCH_SIZE)
test_ds = test_dataset.batch(BATCH_SIZE)
```

Model Data Within TPU Scope

Preserve the input shape for the model:

```
for item in train_ds.take(1):
  s = item[0].shape
in_shape = s[1:]
in_shape
```

We save the input shape to be used later in the model.

Import libraries:

```
from tensorflow.keras.models import Sequential
from tensorflow.keras.layers import Dense, Flatten
```

Create a function to build the model as shown in Listing 5-3.

Listing 5-3. Function to Create the Model

```
def get_model():
  return tf.keras.Sequential([
    Flatten(input_shape=in_shape),
    Dense(256, input_shape=in_shape, activation='relu'),
    Dense(num_classes, activation='softmax')])
```

Create and compile the model within the TPU distribution strategy scope as shown in Listing 5-4.

Listing 5-4. Create and Compile the Model Within TPU Scope

```
with tpu_strategy.scope():
  model = get_model()
  model.compile(
      optimizer='adam',
      loss='sparse_categorical_crossentropy',
      metrics=['accuracy'])
```

To train a model in the Cloud TPU, we must create and compile it inside the TPU strategy scope we created earlier in the chapter.

Inspect the model:

```
model.summary()
```

Train the model:

```
epochs = 60
history = model.fit(train_ds, epochs=epochs,
                    validation_data=(test_ds))
```

We can train a model with as many or as few epochs as we wish. However, large datasets consume memory much faster than small ones. Since digits is very small and images are simple, we can train on more epochs with little worry about memory consumption. Keep in mind that we arrived at 60 epochs through a lot of experimentation. This is why we encourage you to practice a lot!

MNIST Experiment

Although MNIST is being replaced by Fashion-MNIST for deep learning practice, it is still much larger than the digits dataset.

Load the dataset as a TFDS object:

```
import tensorflow_datasets as tfds

train, info = tfds.load(name='mnist', split='train',
                        as_supervised=True, try_gcs=True,
                        with_info=True, shuffle_files=True)
test = tfds.load(name='mnist', split='test',
                        as_supervised=True, try_gcs=True)
```

We leave it to you to explore the metadata since we've already described this dataset object in detail in an earlier chapter.

Build the Input Pipeline

Create a function to scale the data:

```
def scale(image, label):
  image = tf.cast(image, tf.float32) / 255.0
  return image, label
```

Build the pipeline:

```
BATCH_SIZE = 200
SHUFFLE_SIZE = 10000

train_dataset = train.map(scale)\
  .shuffle(SHUFFLE_SIZE).repeat()\
  .batch(BATCH_SIZE).prefetch(1)
test_dataset = test.map(scale)\
  .batch(BATCH_SIZE).prefetch(1)
```

Only shuffle and repeat the training dataset. The *repeat()* method repeats the dataset count *n* number of times. If no number is included, the dataset count is infinite. The advantage of an infinite dataset for training is to avoid the potential last partial batch in each epoch so users don't need to think about scaling the gradients based on the actual batch size. Since the model only learns from the training set, we don't shuffle and repeat the test set.

Model Data Within TPU Scope

Create variables to hold number of images, step size, and validation steps:

```
num_train_img = info.splits['train'].num_examples
num_test_img = info.splits['test'].num_examples
steps_per_epoch = num_train_img // BATCH_SIZE
validation_steps = num_test_img // BATCH_SIZE
```

When repeating data, we must include values for the steps per epoch and validation steps parameters.

Preserve the input shape for the model:

```
for item in train_dataset.take(1):
  s = item[0].shape
mnist_shape = s[1:]
mnist_shape
```

Create a function to build the model as shown in Listing 5-5.

Listing 5-5. Function to Build the Model

```
def create_model():
  return Sequential([
    Flatten(input_shape=mnist_shape),
    Dense(512, activation='relu'),
    Dense(mnist_classes, activation='softmax')
    ])
```

Clear the model and generate a seed:

```
import numpy as np

tf.keras.backend.clear_session()
np.random.seed(0)
tf.random.set_seed(0)
```

Import the library for the loss function:

```
from tensorflow.keras.losses import SparseCategoricalCrossentropy
```

We import the loss function to reduce the characters included in the model. If we didn't do this, we would have to tell the model where the loss function resides. So we would have to include the entire library name. We include this code as an alternative to what we did in previous experiments.

Get the number of classes for the model:

```
mnist_classes = info.features['label'].num_classes
mnist_classes
```

Within the TPU strategy scope, create and compile as shown in Listing 5-6.

Listing 5-6. Create and Compile the Model Within TPU Scope

```
with tpu_strategy.scope():
  model = create_model()
  model.compile(
      optimizer='adam',
      steps_per_execution = 50,
      loss=SparseCategoricalCrossentropy(from_logits=True),
      metrics=['sparse_categorical_accuracy'])
```

Creating the model within the TPUStrategy scope means that we train the model on the TPU system. The idea is to leverage a TPU to speed up learning. In industry, we have to work with systems engineers to take advantage of parallel processing between multiple TPUs and other memory sources. Experiment with *steps_per_execution*. Anything between 2 and *steps_per_epoch* could improve performance.

Train the model:

```
history = model.fit(
    train_dataset, epochs=5,
    steps_per_epoch=steps_per_epoch,
    validation_data=test_dataset,
    validation_steps=validation_steps)
```

Fashion-MNIST Experiment

Since Fashion-MNIST is the current drop-in replacement for MNIST, we demonstrate TPU learning with it.

Fashion-MNIST is a dataset of Zalando's article images consisting of a training set of 60,000 examples and a test set of 10,000 examples. The dataset is intended to serve as a direct drop-in replacement of the original MNIST dataset for benchmarking machine learning algorithms.

Load Fashion-MNIST as a tf.keras dataset:

```
fashion_train, fashion_test = tf.keras.datasets\
  .fashion_mnist.load_data()
```

Transform Datasets into Image and Label Sets

Create image and label sets:

```
train_img, train_lbl = fashion_train
test_img, test_lbl = fashion_test
```

Add the grayscale color dimension to image tensors:

```
fashion_train_img = np.expand_dims(train_img, -1)
fashion_test_img = np.expand_dims(test_img, -1)
```

Inspect image tensors:

```
fashion_train_img.shape, fashion_test_img.shape
```

Create a function to convert tensors to float:

```
def float_it(x):
  return x.astype(np.float32)
```

By converting tensors to NumPy tensors, training performance should improve. Invoke the function:

```
fash_train_img, fash_train_lbl = float_it(
    fashion_train_img), float_it(train_lbl)
fash_test_img, fash_test_lbl = float_it(
    fashion_test_img), float_it(test_lbl)
```

Model Data Within TPU Scope

Import libraries:

```
from tensorflow.keras.models import Sequential
from tensorflow.keras.layers import BatchNormalization,\
  Conv2D, MaxPooling2D, Dropout
```

Create a function to build the model as shown in Listing 5-7.

Listing 5-7. Function to Build the Model

```
def create_model():
  return Sequential([
    BatchNormalization(input_shape=(28,28,1)),
    Conv2D(64, (5, 5), padding='same', activation='elu'),
    MaxPooling2D(pool_size=(2, 2), strides=(2,2)),
    Dropout(0.25),
    BatchNormalization(),
    Conv2D(128, (5, 5), padding='same', activation='elu'),
    MaxPooling2D(pool_size=(2, 2)),
    Dropout(0.25),
    BatchNormalization(),
    Conv2D(256, (5, 5), padding='same', activation='elu'),
    MaxPooling2D(pool_size=(2, 2), strides=(2,2)),
    Dropout(0.25),
    Flatten(),
    Dense(256, activation='elu'),
    Dropout(0.5),
    Dense(10, activation='softmax')
    ])
```

Batch normalization is a technique for training very deep neural networks that standardizes the inputs to a layer for each mini-batch. The technique stabilizes the learning process and dramatically reduces the number of training epochs required to train deep networks.

Clear and seed:

```
tf.keras.backend.clear_session()
np.random.seed(0)
tf.random.set_seed(0)
```

Within the TPU strategy scope, create and compile the model as shown in Listing 5-8.

Listing 5-8. Create and Compile the Model Within TPU Scope

```
with tpu_strategy.scope():
  model = create_model()
  model.compile(
```

```
    optimizer=tf.keras.optimizers.Adam(learning_rate=1e-3, ),
    loss='sparse_categorical_crossentropy',
    metrics=['accuracy'])
```

Train the model:

```
history = model.fit(fash_train_img, fash_train_lbl,
    epochs=17, steps_per_epoch=60,
    validation_data=(fash_test_img, fash_test_lbl),
    validation_freq=17)
```

We add the *validation_freq* parameter as just another option. It is only relevant if validation data is provided. It specifies how many training epochs to run before a new validation run is performed. So we perform a validation every 17 epochs. We include this just to show another parameter within the context of an experiment.

Evaluate the model:

```
loss, acc = model.evaluate(fash_test_img, fash_test_lbl)
print ('loss:', loss)
print ('accuracy:', acc)
```

Save the Trained Model

We can preserve the weights from a trained model:

```
model.save_weights('./fashion_mnist.h5', overwrite=True)
```

Make Inferences

We make inferences (predictions) on this dataset because it is currently the preferred dataset for deep learning competitions. Such competitions are very helpful for students and/or practitioners to gain experience and network with other like-minded people. The previous datasets are great for practice, but are not challenging. That is, it is too easy to get high accuracy (or low loss).

Now that we are done training, let's see how well the model predicts fashion categories!

Get label names:

```
class_labels = ['t_shirt', 'trouser', 'pullover', 'dress',
                'coat', 'sandal', 'shirt', 'sneaker',
                'bag', 'ankle_boots']
```

Create a new model from the saved weights from the trained model:

```
new_model = create_model()
new_model.load_weights('./fashion_mnist.h5')
```

Get 40 prediction arrays from the test set for plotting:

```
preds = new_model.predict(fash_test_img)[:40]
```

Transform prediction arrays into scalar prediction values:

```
pred_40 = [tf.argmax(i).numpy() for i in preds]
```

Since prediction arrays don't identify the actual prediction, we use tf.argmax.

Get images and labels for display as shown in Listing 5-9.

Listing 5-9. Images and Labels for Display

```
images, labels = [], []
for i in range(40):
  img = tf.squeeze(fash_test_img[i])
  images.append(img)
  labels.append(int(fash_test_lbl[i]))
```

Create a function to display predictions as shown in Listing 5-10.

Listing 5-10. Function to Display Prediction Performance

```
def display_test(feature, target, num_images,
                 n_rows, n_cols, cl, p):
  for i in range(num_images):
    plt.subplot(n_rows, 2*n_cols, 2*i+1)
    plt.imshow(feature[i], cmap='nipy_spectral')
    pred = cl[p[i]]
    actual = cl[int(target[i])]
```

```
    title_obj = plt.title(actual + ' (' +\
                          pred + ') ')
    if pred == actual:
      title_obj
    else:
      plt.getp(title_obj, 'text')
      plt.setp(title_obj, color='r')
    plt.tight_layout()
    plt.axis('off')
```

Invoke the function:

```
num_rows, num_cols = 10, 4
num_images = num_rows*num_cols
plt.figure(figsize=(20, 20))
display_test(images, labels, num_images, num_rows,
             num_cols, class_labels, pred_40)
```

Images in red indicate incorrect prediction. The ones in black are correct predictions.

Flowers Experiment

The flowers dataset is not that large, but images are much more complex than those in Fashion-MNIST. Flower images contain many more pixels than the other datasets in this chapter. We believe that it is great practice to work with a complex dataset like flowers.

TPUs are very fast. So the training data streaming into the model must keep pace with the training speed of the model to fully leverage the power of TPUs. The preferred method for TPU usage is to store data into the protobuf-based TFRecord format. The *TFRecord format* stores data in a sequence of binary strings as TFRecords. Binary strings are storage and data transfer efficient.

The flowers dataset is stored on the Google Cloud Storage (GCS) as TFRecords. To fully apply the parallelism that TPUs offer and to avoid bottlenecking on data transfer, we read data as TFRecord files with approximately 230 images per file. The number of 230 is used to equally distribute flowers data into several buckets. We discovered this number through experimentation. We use tf.data.experimental.AUTOTUNE to optimize different parts of input loading. We use AUTOTUNE for this experiment because the dataset contains complex images. This optimization technique wasn't necessary for the earlier experiments.

For a nice tutorial on TPUs in Colab, peruse

https://colab.research.google.com/notebooks/tpu.ipynb

For a nice introduction to TPU data pipeline processing, peruse

https://codelabs.developers.google.com/codelabs/keras-flowers-data

Read Flowers Data as TFRecord Files

Establish the GCS filename pattern:

```
piece1 = 'gs://flowers-public/'
piece2 = 'tfrecords-jpeg-192x192-2/*.tfrec'
TFR_GCS_PATTERN = piece1 + piece2
tfr_filenames = tf.io.gfile.glob(TFR_GCS_PATTERN)
```

Get the number of buckets:

```
num_images = len(tfr_filenames)
print ('Pattern matches {} image buckets.'.format(num_images))
```

Images are contained in 16 buckets (or TFRecord files). We know from earlier flower experiments that there are 3670 images in the flowers dataset. The first 15 TFRecord files contain 230 images each, and the final TFRecord file contains 220 images.

Display all TFRecord files (or buckets):

```
filenames_tfrds = tf.data.Dataset.list_files(TFR_GCS_PATTERN)
for filename in filenames_tfrds.take(16):
  print (filename.numpy())
```

The number of data items (flower images) is written in the name of the TFRecord file. For example, TFRecord file b'gs://flowers-public/tfrecords-jpeg-192x192-2/flowers02-*230*.tfrec' has 230 data items.

Set Parameters for Training

Set parameters for image resizing, pipelining, and epochs:

```
IMAGE_SIZE = [192, 192]
AUTO = tf.data.experimental.AUTOTUNE
BATCH_SIZE = 64
SHUFFLE_SIZE = 100
EPOCHS = 9
```

Set parameters for data splits and labels:

```
VALIDATION_SPLIT = 0.19
CLASSES = ['daisy', 'dandelion', 'roses', 'sunflowers', 'tulips']
```

We use different splits for this experiment because we found this one to perform the best.

Create data splits, validation steps, and steps per epoch as shown in Listing 5-11.

Listing 5-11. Create Data Splits, Validation Steps, and Steps per Epoch

```
split = int(len(tfr_filenames) * VALIDATION_SPLIT)
training_filenames = tfr_filenames[split:]
validation_filenames = tfr_filenames[:split]
print ('Splitting dataset into {} training files and {} '\
        'validation files'\
       .format(len(tfr_filenames), len(training_filenames),
               len(validation_filenames)), end = ' ')
print ('with a batch size of {}.'.format(BATCH_SIZE))

validation_steps = int(3670 // len(tfr_filenames) *\
                       len(validation_filenames)) // BATCH_SIZE
steps_per_epoch = int(3670 // len(tfr_filenames) *\
                      len(training_filenames)) // BATCH_SIZE
print ('There are {} batches per training epoch and {} '\
       'batches per validation run.'\
       .format(BATCH_SIZE, steps_per_epoch, validation_steps))
```

We begin by splitting data into training and test splits. We then display the results of the splits. We create validation steps and steps per epoch for the model. We end by displaying this information.

Create Functions to Load and Process TFRecord Files

Create a function to parse a TFRecord file as shown in Listing 5-12.

Listing 5-12. Function to Parse a TFRecord File

```
def read_tfrecord(example):
  features = {
      'image': tf.io.FixedLenFeature([], tf.string),
      'class': tf.io.FixedLenFeature([], tf.int64)
  }
  example = tf.io.parse_single_example(example, features)
  image = tf.image.decode_jpeg(example['image'], channels=3)
  image = tf.cast(image, tf.float32) / 255.0
  image = tf.reshape(image, [*IMAGE_SIZE, 3])
  class_label = example['class']
  return image, class_label
```

The function accepts an example from a TFRecord file. A dictionary holds datatypes common to TFRecords. The tf.string API converts an image to byte strings (list of bytes). The tf.int64 API converts a class label to a 64-bit integer scalar value. The example is parsed into (image, label) tuples. The image element (a JPEG-encoded image) is decoded into a uint8 image tensor. The image tensor is scaled to the [0, 1] range for faster training. It is then reshaped to a standard size for model consumption. The class label element is cast to a scalar.

Create a function to load TFRecord files as a tf.data.Dataset as shown in Listing 5-13.

Listing 5-13. Function to Load TFRecord Files as a tf.data.Dataset

```
def load_dataset(filenames):
  option_no_order = tf.data.Options()
  option_no_order.experimental_deterministic = False
  dataset = tf.data.TFRecordDataset(
      filenames, num_parallel_reads=AUTO)
```

```
  dataset = dataset.with_options(option_no_order)
  dataset = dataset.map(read_tfrecord, num_parallel_calls=AUTO)
  return dataset
```

The function accepts TFRecord files. For optimal performance, code is included to read from multiple TFRecord files at once. The options setting allows order-altering optimizations. As such, *n* files are read in parallel, and data order is disregarded in favor of reading speed.

Create a function to augment training data:

```
def data_augment(image, label):
  modified = tf.image.random_flip_left_right(image)
  modified = tf.image.random_saturation(modified, 0, 2)
  return modified, label
```

Create a function to build an input pipeline from TFRecord files as shown in Listing 5-14.

Listing 5-14. Build an Input Pipeline from TFRecord Files

```
def get_batched_dataset(filenames, train=False):
  dataset = load_dataset(filenames)
  dataset = dataset.cache()
  if train:
    dataset = dataset.map(data_augment, num_parallel_calls=AUTO)
    dataset = dataset.repeat()
    dataset = dataset.shuffle(SHUFFLE_SIZE)
  dataset = dataset.batch(BATCH_SIZE)
  dataset = dataset.prefetch(AUTO)
  return dataset
```

The function accepts TFRecord files and calls the *load_dataset* function. The function continues by building an input pipeline by caching, repeating, shuffling, batching, and prefetching the dataset. Repeating and shuffling are only mapped to training data. We follow best practices by repeating and shuffling only the training data.

Create Train and Test Sets

Instantiate the datasets:

```
training_dataset = get_batched_dataset(
    training_filenames, train=True)
validation_dataset = get_batched_dataset(
    validation_filenames, train=False)
training_dataset, validation_dataset
```

We build the input pipeline for train and test sets.

Display an image:

```
for img, lbl in training_dataset.take(1):
  plt.axis('off')
  plt.title(CLASSES[lbl[0].numpy()])
  fig = plt.imshow(img[0])
  tfr_flower_shape = img.shape[1:]
```

We display an image and preserve the image shape for the model.

Model Data

Clear and seed:

```
tf.keras.backend.clear_session()
np.random.seed(0)
tf.random.set_seed(0)
```

Create a function to build the model as shown in Listing 5-15.

Listing 5-15. Function to Create the Model

```
def create_model():
  return Sequential([
    Conv2D(32, (3, 3), activation = 'relu'),
    MaxPooling2D(2, 2),
    Conv2D(64, (3, 3), activation='relu'),
    MaxPooling2D(2, 2),
```

```
    Conv2D(128, (3, 3), activation='relu'),
    MaxPooling2D(2),
    Conv2D(128, (3, 3), activation='relu'),
    MaxPooling2D(2, 2),
    Flatten(),
    Dense(512, activation='relu'),
    Dense(num_classes, activation='sigmoid')
])
```

Within the TPU strategy scope, create and compile the model as shown in Listing 5-16.

Listing 5-16. Create and Compile the Model Within TPU Scope

```
with tpu_strategy.scope():
  flower_model = create_model()
  flower_model.compile(
      optimizer='adam',
      loss=tf.losses.SparseCategoricalCrossentropy(),
      metrics=['accuracy'])
```

Train the model:

```
history = flower_model.fit(training_dataset, epochs=EPOCHS,
                    verbose=1, steps_per_epoch=steps_per_epoch,
                    validation_steps=validation_steps,
                    validation_data=validation_dataset)
```

Make Inferences

Let's see how well the model predicts flower categories!

Grab 40 predictions from the validation (test) set:

```
preds = flower_model.predict(validation_dataset)[:40]
```

Convert prediction arrays to scalar prediction values:

```
pred_40 = [tf.argmax(i).numpy() for i in preds]
```

Create images and labels for display:

```
images, labels = [], []
for img, lbl in validation_dataset.take(1):
  for i in range(40):
    actual_img = tf.squeeze(img[i])
    images.append(actual_img)
    labels.append(lbl[i].numpy())
```

Create a function to display predictions as shown in Listing 5-17.

Listing 5-17. Function to Display Predictions

```
def display_test(feature, target, num_images,
                 n_rows, n_cols, cl, p):
  for i in range(num_images):
    plt.subplot(n_rows, 2*n_cols, 2*i+1)
    plt.imshow(feature[i])
    pred = cl[p[i]]
    actual = cl[int(target[i])]
    title_obj = plt.title(actual + ' (' +\
                          pred + ') ')
    if pred == actual:
      title_obj
    else:
      plt.getp(title_obj, 'text')
      plt.setp(title_obj, color='r')
    plt.tight_layout()
    plt.axis('off')
```

Invoke the function:

```
num_rows, num_cols = 10, 4
num_images = num_rows*num_cols
plt.figure(figsize=(20, 20))
display_test(images, labels, num_images, num_rows,
             num_cols, CLASSES, pred_40)
```

CHAPTER 6

Simple Transfer Learning with TensorFlow Hub

Transfer learning is the process of creating new learning models by fine-tuning previously trained neural networks. Instead of training a network from scratch, we download a pre-trained open source learning model and fine-tune it for our own purpose. A *pre-trained model* is one that is created by someone else to solve a similar problem. We can use one of these instead of building our own model. A big advantage is that a pre-trained model has been crafted by experts, so we can be confident that it performs at a high level (in most cases). Another advantage is that we don't have to have a lot of data to use a pre-trained model.

In transfer learning, a machine exploits the knowledge gained from a previous task to improve generalization about another. For example, in training a classifier to predict whether an image contains food, we can use the knowledge it gained during training to recognize drinks. So transfer learning can save time, provides better neural network performance in most cases, and doesn't require a lot of data.

TensorFlow Hub is a repository of trained machine learning models that are easily integrated into deep learning experiments. We demonstrate simple transfer learning with TensorFlow Hub code examples.

Image classification models can have millions of parameters. Training them from scratch requires a lot of labeled training data and a lot of computing power. With transfer learning, we don't have to train from scratch! We can take a piece of a model that has already been trained on a related task and reuse it in a new model. Since we use the weights from a pre-trained model, training time is drastically reduced! So we don't have to train a gigantic neural net because it's already been trained! And training time is reduced because we use the pre-trained model on our dataset.

Notebooks for chapters are located at the following URL:

https://github.com/paperd/deep-learning-models

D. Paper, *State-of-the-Art Deep Learning Models in TensorFlow*, https://doi.org/10.1007/978-1-4842-7341-8_6

Pre-trained Models for Transfer Learning

If we don't have enough training data, it is often a good idea to reuse the lower layers of a pre-trained model. Reusing lower layers of a pre-trained model is commonly referred to as transfer learning. We only train with the lower layers and leave the others frozen. *Lower layers* refer to the general (or problem-independent) features of a model. The higher layers refer to specific (or problem-dependent) features of a model. The final layers of a pre-trained model are also referred to as the final layers.

The idea is to create new deep learning models by fine-tuning previously trained ones for our own purpose. Simply, import the trained weights of a pre-trained model and change the final layer (or several layers) of the model. We then retrain those layers on our own dataset. We save training time and get respectable performance.

We demonstrate how to use pre-trained models with examples. We begin with an experiment using MobileNet-v2 and end with an experiment using Inception-v3. Each pre-trained model is explained in its corresponding sections of the chapter.

Import the TensorFlow Library

Import the library and alias it as **tf**:

```
import tensorflow as tf
```

GPU Hardware Accelerator

As a convenience, we provide the steps to enable the GPU in a Colab notebook:

1. Click *Runtime* in the top-left menu.
2. Click *Change runtime type* from the drop-down menu.
3. Choose *GPU* from the *Hardware accelerator* drop-down menu.
4. Click *Save.*

Verify that the GPU is active:

```
tf.__version__, tf.test.gpu_device_name()
```

If '/device:GPU:0' is displayed, the GPU is active. If '..' is displayed, the regular CPU is active.

Note If you get the error **NAME 'TF' IS NOT DEFINED**, re-execute the code to import the TensorFlow library!

TensorFlow Hub

We model flowers data with pre-trained TensorFlow SavedModels from TensorFlow Hub for image feature extraction. A *SavedModel* is a directory containing serialized signatures and the state needed to run them including variable values and vocabularies. *TensorFlow Hub* is a repository of pre-trained TensorFlow models. The pre-trained models were trained on very large and general datasets.

For more information on TensorFlow Hub, peruse

https://tfhub.dev/

For tutorials on TensorFlow Hub, peruse

www.tensorflow.org/hub/tutorials

We use two pre-trained TensorFlow Hub models for transfer learning. We begin with the MobileNet-v2 pre-trained model. We then use the Inception-v3 pre-trained model and compare results between the two.

MobileNet-v2

MobileNet-v2 is a convolutional neural network that is 53 layers deep. The pre-trained version of the network is trained on 1.4 million images and 1,000 classes of web images from the ImageNet database. The model operates on 224 × 224 pixel images. The default training batch size is 1024, which means that each iteration operates on 1024 of those images.

The *ImageNet project* is a large visual database designed for use in visual object recognition software research. More than 14 million images have been hand-annotated by the project to indicate what objects are pictured.

Flowers MobileNet-v2 Experiment

The first experiment uses the MobileNet-v2 pre-trained model to classify flowers.

Load Flowers as a TFDS Object

Load the TFDS object with a 75% split for the training set, 15% split for the validation set, and 10% split for the test set:

```
import tensorflow_datasets as tfds

(test, valid, train), info = tfds.load(
    'tf_flowers', as_supervised=True,
    split = ['train[:10%]', 'train[10%:25%]', 'train[25%:]'],
    with_info=True, try_gcs=True)
```

Explore Metadata

Display general information with the info object:

```
info
```

Display the number of examples in the data splits:

```
num_train_img = info.splits['train[25%:]'].num_examples
num_valid_img = info.splits['train[10%:25%]'].num_examples
num_test_img = info.splits['train[:10%]'].num_examples
print ('train images:', num_train_img)
print ('valid images:', num_valid_img)
print ('test images:', num_test_img)
```

We like to display this information for each experiment just to make sure that we are splitting the data as intended.

Manually calculate the number of examples in the data splits as shown in Listing 6-1.

Listing 6-1. Manually Verify Examples in Each Split

```
num_train_examples = 0
num_valid_examples = 0
num_test_examples = 0
```

```
for example in train:
  num_train_examples += 1

for example in valid:
  num_valid_examples += 1

for example in test:
  num_test_examples += 1

print('Total Number of Training Images: {}'\
      .format(num_train_examples))
print('Total Number of Validation Images: {}'\
      .format(num_valid_examples))
print('Total Number of Testing Images: {}'\
      .format(num_test_examples))
```

The code in Listing 6-1 is optional. We just want to show you how to manually check sizes of the learning sets.

Get labels and number of classes:

```
class_labels = info.features['label'].names
num_classes = info.features['label'].num_classes
class_labels, num_classes
```

Display Images and Shapes

Display some images from the dataset:

```
fig = tfds.show_examples(train, info)
```

Display image shapes:

```
for i, example in enumerate(train.take(5)):
  print('Image {} shape: {} label: {}'\
        .format(i+1, example[0].shape,
               example[1]))
```

Images in the flowers dataset are not all the same size. So we must resize images to a standard size to make them consumable by TensorFlow models.

Build the Input Pipeline

Create a function to reformat images:

```
def format_image(image, label):
  image = tf.image.resize(image, (224, 224)) /255.0
  return image, label
```

The function resizes and scales an image. The resolution expected by MobileNet-v2 is (224, 224). The function takes an "image" and a "label" as arguments and returns the new "image" and corresponding "label" in the desired form.

Map the function to training, validation, and test examples. And apply other transformations:

```
BATCH_SIZE = 367

train_batches = train.shuffle(num_train_img//4).\
  map(format_image).batch(BATCH_SIZE).prefetch(1)

validation_batches = valid.map(format_image).\
  batch(BATCH_SIZE).prefetch(1)

test_batches = test.map(format_image).\
  batch(BATCH_SIZE).prefetch(1)
```

Fortuitously, we found that the batch size of 367 worked extremely well! We tried the default batch size of 1024 and displayed labels based on this batch size. Surprisingly, we got an error that said something like *our index was out of bounds for a size of 367*. So we changed the batch size to 367 and noticed that the model performed better in terms of increased validation accuracy than with the batch size of 1024. Moreover, training time was reduced.

Create a Feature Vector

A **feature** is an individual measurable property or characteristic of a phenomenon being observed. A **feature vector** is a vector containing multiple elements about an object. A feature is represented by a feature vector. Combining feature vectors together creates a feature space. A **feature space** is the vector space associated with feature vectors. A feature may represent just a single pixel or an entire image. In our case, a feature represents an entire image depicted by a feature vector of pixels.

Feature extraction is the process of gleaning a subset of the initial features from a dataset. The extracted features are expected to contain relevant information from the input data. The goal of feature extraction is to perform a desired task by using a reduced representation (or subset) instead of the complete initial data. Choosing informative, discriminating, and independent features is a crucial step for effective algorithms in pattern recognition, classification, and regression.

Create a feature extractor with the pre-trained model:

```
import tensorflow_hub as hub

piece1 = 'https://tfhub.dev/google/tf2-preview/'
piece2 = 'mobilenet_v2/feature_vector/4'
URL = piece1 + piece2
feature_extractor_mn = hub.KerasLayer(
    URL, input_shape=(224, 224, 3))
```

Create a feature extractor using the MobileNet-v2 feature vector model. The feature extractor is the partial model from TensorFlow Hub (without the final classification layer).

Freeze the Pre-trained Model

Freeze the variables in the feature extractor layer so that training only modifies the final classifier layer:

```
feature_extractor_mn.trainable = False
```

The result of not freezing the pre-trained layers destroys the information they contain during future training rounds.

Create a Classification Head

Create a classification head to leverage the pre-trained model for the dataset. The classification head consists of a simple sequential model that includes the pre-trained model and the new classification layer.

Import libraries:

```
from tensorflow.keras.models import Sequential
from tensorflow.keras.layers import Dense, Dropout
```

Clear and seed:

```
import numpy as np

tf.keras.backend.clear_session()
np.random.seed(0)
tf.random.set_seed(0)
```

Create the classification head:

```
mobile_model = tf.keras.Sequential([
  feature_extractor_mn,
  Dropout(0.5),
  Dense(num_classes)])
```

A *classification head* is just a container that holds the pre-trained model and any additions that we might wish to add. Notice that the first layer of the model we just created is the feature extractor. Also notice that the model is **extremely** simple because we leverage the pre-trained weights from MobileNet-v2!

Note For our experiment, the classification head is the new model we create that contains the pre-trained model (in this case the pre-trained model is a feature extractor), a Dropout layer (to reduce overfitting), and a Dense layer (to enable classification). To implement a pre-trained model, a mechanism is required, and it is referred to as a classification head.

Compile and Train the Model

Compile:

```
from tensorflow.keras.losses import SparseCategoricalCrossentropy

mobile_model.compile(
  optimizer='adam',
  loss=SparseCategoricalCrossentropy(from_logits=True),
  metrics=['accuracy'])
```

Train:

```
EPOCHS = 6

history = mobile_model.fit(
    train_batches, epochs=EPOCHS,
    validation_data=validation_batches)
```

Although our model is very simple with no tuning, we get pretty good accuracy with just six epochs because MobileNet-v2 was carefully designed over a long time by experts and then trained on the massive ImageNet dataset. Also, training time is very reasonable.

Note With a very large model like MobileNet-v2, training time would be significant if we didn't use its pre-trained weights.

Visualize Performance

Plot validation accuracy and loss as shown in Listing 6-2.

Listing 6-2. Visual Accuracy and Loss Performance

```
import matplotlib.pyplot as plt

acc = history.history['accuracy']
val_acc = history.history['val_accuracy']

loss = history.history['loss']
val_loss = history.history['val_loss']

epochs_range = range(EPOCHS)

plt.figure(figsize=(8, 8))
plt.subplot(1, 2, 1)
plt.plot(epochs_range, acc, label='Training Accuracy')
plt.plot(epochs_range, val_acc, label='Validation Accuracy')
plt.legend(loc='lower right')
plt.title('Training and Validation Accuracy')
```

```
plt.subplot(1, 2, 2)
plt.plot(epochs_range, loss, label='Training Loss')
plt.plot(epochs_range, val_loss, label='Validation Loss')
plt.legend(loc='upper right')
plt.title('Training and Validation Loss')
plt.show()
```

We find it amazing that we can get respectable results from a pre-trained model on a dataset upon which it wasn't trained.

Make Predictions from Test Data

We predict on test data because the model has **never** seen it!

Predict on test data:

```
predictions = mobile_model.predict(test_batches)
```

Display class labels:

```
class_labels
```

Inspect the First Prediction

We inspect a prediction (or inference) from the dataset to show you the raw values returned from the *predict* method. Since inference is the main goal of classification learning, we find it valuable to show you what is happening underneath the hood (so to speak).

Get the first prediction array:

```
predictions[0]
```

The returned array is the raw prediction.

Use the *np.argmax()* function to get the prediction, which is the value with the highest probability from the prediction array for the first image:

```
predicted_id = np.argmax(predictions[0])
predicted_id
```

Convert the label to its class name:

```
class_labels[predicted_id]
```

Get the labels from the first batch:

```
for img, lbl in test_batches.take(1):
  print (lbl)
```

The number of labels matches the batch size.

Get the first label:

```
class_labels[lbl[0].numpy()]
```

The prediction was correct if the first label matches the prediction for the first image!

Inspect the First Batch of Predictions

Alternatively, we can convert *test_batches* to an iterator:

```
image_batch, label_batch = next(iter(test_batches))

images = image_batch.numpy()
labels = label_batch.numpy()

class_labels[labels[0]]
```

Get the first batch from the iterator, convert images and labels to NumPy, and display the first label.

Display labels from the first batch:

```
labels
```

Convert the batch of labels to named labels:

```
named_labels = [class_labels[labels[i]]
                for i, lbl in enumerate(range(BATCH_SIZE))]
named_labels
```

Get predictions from the first batch:

```
predicted_batch = [np.argmax(predictions[i])
                   for i, _ in enumerate(range(BATCH_SIZE))]
predicted_batch
```

Convert predictions to named predictions:

```
named_pred = [class_labels[predicted_batch[i]]
              for i, lbl in enumerate(range(BATCH_SIZE))]
named_pred
```

Plot Predictions

The visualization shows actual images from the first test batch. If the prediction is correct, the title is blue. If not, the title is red. If the prediction is incorrect, the prediction is displayed along with the actual label in parentheses.

Display the predictions as shown in Listing 6-3.

Listing 6-3. Prediction Plot

```
plt.figure(figsize=(20,20))
for n in range(30):
  plt.subplot(6,5,n+1)
  plt.subplots_adjust(hspace = 0.3)
  plt.imshow(images[n])
  color = 'blue' if labels[n] == predicted_batch[n] else 'red'
  if labels[n] != predicted_batch[n]:
    t = named_pred[n].title() +\
        ' (' +named_labels[n].title() + ')'
  else:
    t = named_pred[n].title()
  plt.title(t, color=color)
  plt.axis('off')
  st = 'Model predictions (blue: correct, red: incorrect)'
_ = plt.suptitle(st)
```

Flowers Inception-v3 Experiment

For the second experiment, we use the Inception-v3 pre-trained model to classify flowers. *Inception-v3* is a 48-layer-deep convolutional neural network used for image recognition that has been shown to attain greater than 78.1% accuracy on the ImageNet dataset. The pre-trained model operates on 299 × 299 images. Default training batch size is 1024, which means that each iteration operates on 1024 of those images.

For an advanced guide to Inception-v3 on Cloud TPU, peruse *https://cloud.google.com/tpu/docs/inception-v3-advanced*

Build the Input Pipeline

Recreate the function to reformat images:

```
def format_image(image, label):
  image = tf.image.resize(image, (299, 299)) / 255.0
  return image, label
```

The function resizes and scales an image. The resolution expected by Inception-v3 is (299, 299). The function takes an "image" and a "label" as arguments and returns the new "image" and corresponding "label" in the desired form.

Build the input pipeline for Inception-v3:

```
BATCH_SIZE = 367

train_im = train.shuffle(num_train_img//4).\
  map(format_image).batch(BATCH_SIZE).prefetch(1)

validation_im = valid.map(format_image).\
  batch(BATCH_SIZE).prefetch(1)

test_im = test.map(format_image).\
  batch(BATCH_SIZE).prefetch(1)
```

Map the function to training, validation, and test examples. And apply other transformations.

Create a feature extractor:

```
piece1 = 'https://tfhub.dev/google/tf2-preview/'
piece2 = 'inception_v3/feature_vector/4'
URL = piece1 + piece2
feature_extractor_im = hub.KerasLayer(URL,
  input_shape=(299, 299, 3),
  trainable=False)
```

Freeze the pre-trained model:

```
feature_extractor_im.trainable = False
```

Build the Model

Clear and seed:

```
tf.keras.backend.clear_session()
np.random.seed(0)
tf.random.set_seed(0)
```

Create the model:

```
inception_model = tf.keras.Sequential([
  feature_extractor_im,
  Dropout(0.5),
  Dense(num_classes)])
```

We just substitute the Inception-v3 feature extractor to get its pre-trained weights!

Compile and Train

Compile:

```
inception_model.compile(
  optimizer='adam',
  loss=SparseCategoricalCrossentropy(from_logits=True),
  metrics=['accuracy'])
```

Train:

```
EPOCHS = 6

history = inception_model.fit(
    train_im, epochs=EPOCHS,
    validation_data=validation_im)
```

Visualize Performance

Visualize training loss and accuracy as shown in Listing 6-4.

Listing 6-4. Training Performance

```
import matplotlib.pyplot as plt

acc = history.history['accuracy']
val_acc = history.history['val_accuracy']

loss = history.history['loss']
val_loss = history.history['val_loss']

epochs_range = range(EPOCHS)

plt.figure(figsize=(8, 8))
plt.subplot(1, 2, 1)
plt.plot(epochs_range, acc, label='Training Accuracy')
plt.plot(epochs_range, val_acc, label='Validation Accuracy')
plt.legend(loc='lower right')
plt.title('Training and Validation Accuracy')

plt.subplot(1, 2, 2)
plt.plot(epochs_range, loss, label='Training Loss')
plt.plot(epochs_range, val_loss, label='Validation Loss')
plt.legend(loc='upper right')
plt.title('Training and Validation Loss')
plt.show()
```

Note We are not really concerned about comparing model performance between Inception and MobileNet because results are always going to be tied to the data that you want to model. That is, one might work better on one dataset but not as well on another. This is why we show you how to implement both models.

Predictions

Make predictions:

```
im_predictions = inception_model.predict(test_im)
```

Get a batch of predictions and convert them to named predictions:

```
im_pred_batch = [np.argmax(im_predictions[i])
                 for i, _ in enumerate(range(BATCH_SIZE))]
im_named_pred = [class_labels[im_pred_batch[i]]
                 for i, lbl in enumerate(range(BATCH_SIZE))]
```

Grab the first batch of images and labels from the test set:

```
im_image_batch, im_label_batch = next(iter(test_im))

im_images = im_image_batch.numpy()
im_labels = im_label_batch.numpy()
```

Convert the labels to named labels:

```
im_named_labels = [class_labels[im_labels[i]]
                   for i, lbl in enumerate(range(BATCH_SIZE))]
```

Plot Model Predictions

Create a plot function as shown in Listing 6-5.

Listing 6-5. Plot Function

```
def plot_pred(images, labels, named_labels, named_pred):
  plt.figure(figsize=(20,20))
  for n in range(30):
    plt.subplot(6,5,n+1)
    plt.subplots_adjust(hspace = 0.3)
    plt.imshow(images[n])
    color = 'blue' if named_labels[n] == named_pred[n] else 'red'
    if named_labels[n] != named_pred[n]:
      t = named_pred[n].title() +\
      ' (' +named_labels[n].title() + ')'
```

```
    else:
      t = named_pred[n].title()
    plt.title(t, color=color)
    plt.axis('off')
    st = 'Model predictions (blue: correct, red: incorrect)'
    _ = plt.suptitle(st)
```

Wrap plotting logic in a function.

Invoke the function:

```
plot_pred(im_images, im_labels, im_named_labels, im_named_pred)
```

Summary

The performance of MobileNet-v2 and Inception-v3 was quite similar with the flowers dataset. Both models showed no overfitting. Accuracy for both models was above 80%. Of course, we only trained for six epochs. But we can't predict how well a pre-trained model works on a given dataset. The idea of pre-trained models is to save you time and energy. And this is why there are so many available pre-trained models. We speculate that there are going to be many new models created in the near future.

CHAPTER 7

Advanced Transfer Learning

We introduce advanced transfer learning with code examples based on several transfer learning architectures. The code examples train learning models with these architectures.

Notebooks for chapters are located at the following URL:

https://github.com/paperd/deep-learning-models

Transfer Learning

Transfer learning is based on the idea that the feature a network learns for a problem can be reused for a variety of other tasks. In ML and pattern recognition, a **feature** is an individual measurable property or characteristic of a phenomenon being observed. So effective learning algorithms are able to glean informative, discriminating, and independent features from data.

Transfer learning models are amazing because they can reuse the features they've learned on new data! Leveraging transfer learning saves the time it takes to create a new model, test it, and tweak it until it provides the desired results. Also, transfer learning models are created by experienced data scientists who have spent years tweaking and perfecting available transfer learning models.

Transfer learning models can boost accuracy without taking much time to converge when compared to a model trained from scratch. But this doesn't mean that a pre-trained model is clearly better than one created from scratch. A ML model reaches convergence when it achieves a state during training where loss settles to within an error range around the final value. So a model converges when additional training doesn't improve the model.

D. Paper, *State-of-the-Art Deep Learning Models in TensorFlow*, https://doi.org/10.1007/978-1-4842-7341-8_7

In a recent conversation with a practicing data scientist, he informed us that he creates some of his own models from scratch. On a current project, he is using an existing (pre-trained) model. For other projects, he builds his own models. It just depends on the task. He also let us know that he is not as concerned with speed as he is with convergence. Keep in mind that this person has a PhD in data science with many years of industry experience.

If a pre-trained neural network is effective, the features it learns can be used for other tasks. When humans learn how to perform a new task, we seldom start from scratch. We carry over all that we have learned in our lifetime to quickly learn new stuff. We can often learn from a single training example. But other times it actually hinders our development. Of course, babies don't learn this way because they don't have the same level of prior knowledge.

We demonstrated in the previous chapter that transfer learning can be used on datasets that it has never seen. In this chapter, we present a newer pre-trained model to show how it works with different datasets. We also show you how to directly implement transfer learning without using the TensorFlow Hub library. Implementing transfer learning directly provides more flexibility as we demonstrate with our experiments.

We present four transfer learning experiments. We begin with the beans experiment. We continue with the Stanford Dogs, flowers, and rock-paper-scissors experiments. We strongly believe that practice makes perfect. Code tends to be similar for each experiment, but each dataset is treated differently.

With all of our experiments, we set up the Colab ecosystem. So begin by importing the main TensorFlow library and instantiating the GPU.

Import the TensorFlow Library

Import the library and alias it as **tf**:

```
import tensorflow as tf
```

Aliasing the TensorFlow library as tf is common practice.

GPU Hardware Accelerator

As a convenience, we provide the steps to enable the GPU in a Colab notebook:

1. Click *Runtime* in the top-left menu.
2. Click *Change runtime type* from the drop-down menu.
3. Choose *GPU* from the *Hardware accelerator* drop-down menu.
4. Click *Save*.

Verify that the GPU is active:

```
tf.__version__, tf.test.gpu_device_name()
```

If '/device:GPU:0' is displayed, the GPU is active. If '..' is displayed, the regular CPU is active.

Note If you get the error **NAME 'TF' IS NOT DEFINED**, re-execute the code to import the TensorFlow library!

Beans Experiment

Beans is a TensorFlow dataset (TFDS) of bean plant images taken in the field with smartphone cameras. It consists of three classes (bean_rust, angular_leaf_spot, healthy). Two of the three classes are *angular leaf spot* and *bean rust*, which are diseases that can befell bean plants. The third class is *healthy*. So a bean plant in this dataset is either healthy or afflicted with one of the two diseases. Data was annotated by experts from the National Crops Resources Research Institute (NaCRRI) in Uganda and collected by the Makerere AI research lab.

We train on beans data with two pre-trained models. One of the models is new to you, and the other is Inception that was demonstrated in the previous chapter. We show both for contrast. Even if one works better than the other on this dataset, there is no guarantee that it will do so on another! But using pre-trained models is pretty straightforward once you get used to working with them. So trying different ones is probably a good strategy.

Load Beans

Load beans as a TFDS from the Google Cloud Service (GCS):

```
import tensorflow_datasets as tfds

beans, beans_info = tfds.load(
    'beans', with_info=True, as_supervised=True,
    try_gcs=True)
```

Although not required, we recommend loading a TFDS from GCS.

Explore the Data

Display the metadata:

```
beans_info
```

For simplicity, assign the splits to variables:

```
train = beans['train']
valid = beans['validation']
test = beans['test']
```

From the metadata, we know that training data has 1,034 examples, validation data has 133 examples, and test data has 128 examples.

Get labels and number of classes:

```
class_labels = beans_info.features['label'].names
num_classes = beans_info.features['label'].num_classes
class_labels, num_classes
```

Check image sizes:

```
for img, lbl in train.take(10):
  print (img.shape)
```

From the sample, it appears that all images are 500 × 500 × 3.

Visualize

Visualize with *show_examples*:

```
fig = tfds.show_examples(train, beans_info)
```

Reformat Images

Resize and process images for the Xception model:

```
def preprocess(image, label):
  resized_image = tf.image.resize(image, [224, 224])
  final_image = tf.keras.applications.xception.\
               preprocess_input(resized_image)
  return final_image, label
```

We resize images to 224 × 224, which is the expected size of Xception images. We preprocess resized images with the Xception preprocessing API and return the final image and label. The function is used in the next section.

Build the Input Pipeline

Transform training, validation, and test data to TensorFlow consumable objects:

```
BATCH_SIZE = 32
shuffle = 250

train_ds = train.shuffle(shuffle).\
  map(preprocess).batch(BATCH_SIZE).prefetch(1)
valid_ds = valid.map(preprocess).batch(BATCH_SIZE).prefetch(1)
test_ds = test.map(preprocess).batch(BATCH_SIZE).prefetch(1)
```

Visualize as shown in Listing 7-1.

Listing 7-1. Visualize Training Examples

```
import matplotlib.pyplot as plt

plt.figure(figsize=(12, 12))
for img, lbl in train_ds.take(1):
```

```
for index in range(9):
  plt.subplot(3, 3, index + 1)
  plt.imshow(img[index] / 2 + 0.5)
  plt.title(class_labels[lbl[index]])
  plt.axis('off')
```

We provide different visualization code for the task at hand. Feel free to create your own visualization code to gain experience with Python data science tasks.

Model with Xception

We introduce the Xception pre-trained model to broaden your experience. Also, Xception is the most recent one introduced to the public.

Xception is one of six state-of-the-art image classifiers pre-trained on the ImageNet dataset. The other five are MobileNet, VGG16, VGG19, ResNet50, and Inception-v3. Xception slightly outperforms Inception-v3 on the ImageNet dataset and vastly outperforms it on a larger image classification dataset with 17,000 classes (or more). Most importantly, it has the same number of model parameters as Inception implying a greater computational efficiency. VGG16 and VGG19 have smaller convolution layers but take much longer to train than Xception. ResNet model size is comparable, but Xception trains faster and produces better training results.

The Xception model was proposed by Francois Chollet in 2017. *Xception* is an extension of the Inception architecture that replaces the standard Inception modules with depthwise Separable Convolutions. Xception is pre-trained on ImageNet. Xception often outperforms VGGNet, ResNet, and Inception-v3 models. As a side note, Chollet is also the author of Keras.

Create a Model

Clear and seed:

```
import numpy as np

tf.keras.backend.clear_session()
np.random.seed(0)
tf.random.set_seed(0)
```

Create a base model with Xception:

```
Xception = tf.keras.applications.xception.Xception
xception_model = Xception(
    weights='imagenet', include_top=False)
```

Exclude the top layer of the network by setting *include_top=False*. So the global average pooling layer and the Dense output layer are excluded from Xception, which means that we must include both layers in the classification head.

We exclude the top layer to train faster. However, training with all layers may increase learning. We show you how to do this later in the chapter.

Explore base model layers:

```
tf.keras.utils.plot_model(
    xception_model,
    show_shapes=True,
    show_layer_names=True)
```

The tf.keras.utils.plot_model API provides a detailed graphical description of the Xception model.

Get the number of layers:

```
len(xception_model.layers)
```

Xception has 132 layers.

Import libraries:

```
from tensorflow.keras.models import Sequential
from tensorflow.keras.layers import Dense, Dropout,\
                                    GlobalAveragePooling2D
```

Build the final model (classification head):

```
x_model = tf.keras.Sequential([
  xception_model,
  GlobalAveragePooling2D(),
  Dropout(0.5),
  Dense(num_classes, activation='softmax')])
```

We excluded the top layer of the pre-trained network, which has a global average pooling layer and a Dense output layer. So we must add our own global average pooling layer and a Dense output layer with three classes and softmax activation.

Get the layout of the final model:

```
tf.keras.utils.plot_model(
    x_model,
    show_shapes=True,
    show_layer_names=True)
```

Notice the simplicity of our final model!

Model the Data

Freeze the weights of the pre-trained layers:

```
for layer in xception_model.layers:
  layer.trainable = False
```

We add the preceding step to inform the compiler our intention of freezing the top layer.

Compile:

```
optimizer = tf.keras.optimizers.SGD(
    lr=0.2, momentum=0.9, decay=0.01)

x_model.compile(
    loss='sparse_categorical_crossentropy',
    optimizer=optimizer,
    metrics=['accuracy'])
```

Train:

```
history = x_model.fit(
    train_ds, validation_data=valid_ds, epochs=10)
```

Notice that training time is minimal.

Create a function to visualize training performance as shown in Listing 7-2.

Listing 7-2. Visualization Function

```
def visualize(span):
  acc = history.history['accuracy']
  val_acc = history.history['val_accuracy']
  loss = history.history['loss']
  val_loss = history.history['val_loss']
  epochs_range = span
  plt.figure(figsize=(8, 8))
  plt.subplot(1, 2, 1)
  plt.plot(epochs_range, acc, label='Training Accuracy')
  plt.plot(epochs_range, val_acc, label='Validation Accuracy')
  plt.legend(loc='lower right')
  plt.title('Training and Validation Accuracy')
  plt.subplot(1, 2, 2)
  plt.plot(epochs_range, loss, label='Training Loss')
  plt.plot(epochs_range, val_loss, label='Validation Loss')
  plt.legend(loc='upper right')
  plt.title('Training and Validation Loss')
  plt.show()
```

Invoke:

```
visualize(range(10))
```

Note Be sure to feed the visualize function a range equal to the number of epochs upon which the model is trained.

We set a very aggressive learning rate, but still got respectable performance. Experiment with learning rates to see if you can increase performance.

Note Setting learning rate to a high value generally enables a model to learn faster. However, the cost can be a suboptimal final set of weights. Setting learning rate to a low value may allow a model to learn a more optimal or even globally optimal set of weights, but may take significantly longer to train.

Model Trained Data with Unfrozen Layers

Validation accuracy is pretty good, but doesn't get better. So the top layers are pretty well trained. That is, accuracy reaches a plateau.

Now we are ready unfreeze all the layers and continue training as shown in Listing 7-3.

Listing 7-3. Continue Training with All Layers Unfrozen

```
for layer in xception_model.layers:
  layer.trainable = True

optimizer = tf.keras.optimizers.SGD(
    learning_rate=0.01, momentum=0.9,
    nesterov=True, decay=0.001)

x_model.compile(
    loss='sparse_categorical_crossentropy',
    optimizer=optimizer, metrics=['accuracy'])

history = x_model.fit(
    train_ds, validation_data=valid_ds, epochs=10)
```

Intially, a good strategy is to train with top layers frozen for speed. Setting a high learning rate improves training speed even more. Once validation accuracy plateaus, we know that the top layers are trained. We can then continue training by unfreezing the top layers and setting a lower learning rate. We use a *lower learning rate* to avoid damaging the pre-trained weights.

We still train on the same model until we clear the model sessions. This is why we clear model sessions before we train a new model.

Visualize:

```
visualize(range(10))
```

Model Beans with Inception

Let's see how Inception compares to Xception. We show in this section how easy it is to interchange pre-trained models.

Inception-v3 is a pre-trained convolutional neural network model that is 48 layers deep. It is a version of the network already trained on more than a million images from the ImageNet database. The pre-trained network can classify images into 1000 object categories such as keyboard, mouse, pencil, and many animals.

Create a Model

Clear and seed:

```
tf.keras.backend.clear_session()
np.random.seed(0)
tf.random.set_seed(0)
```

Create the base model:

```
inception_v3 = tf.keras.applications.InceptionV3
inception_model = inception_v3(
    include_top=False, weights='imagenet',
    input_shape=(224, 224, 3))
```

Explore base model layers:

```
tf.keras.utils.plot_model(
    inception_model,
    show_shapes=True,
    show_layer_names=True)
```

Get number of layers:

```
len(inception_model.layers)
```

Whew! Inception is extremely large (311 layers) and complex!

Create the final model:

```
i_model = tf.keras.Sequential([
  inception_model,
  GlobalAveragePooling2D(),
  Dropout(0.5),
  Dense(num_classes, activation='softmax')])
```

We can use shape (224, 224, 3) since include_top is False. Otherwise, the input shape must be (299, 299, 3) for Inception.

Model the Data

Freeze the weights of the pre-trained layers:

```
for layer in inception_model.layers:
  layer.trainable = False
```

Compile:

```
optimizer = tf.keras.optimizers.RMSprop(lr=0.1)

i_model.compile(
    loss='sparse_categorical_crossentropy',
    optimizer=optimizer, metrics=['accuracy'])
```

Train:

```
history = i_model.fit(
    train_ds, validation_data=valid_ds, epochs=10)
```

Visualize:

```
visualize(range(10))
```

Model Trained Data with Unfrozen Layers

Unfreeze all the layers and continue training as shown in Listing 7-4.

Listing 7-4. Continue Training with All Layers Unfrozen

```
for layer in inception_model.layers:
  layer.trainable = True

optimizer = tf.keras.optimizers.RMSprop(lr=0.0001)

i_model.compile(
    loss='sparse_categorical_crossentropy',
    optimizer=optimizer, metrics=['accuracy'])
```

```
history = i_model.fit(
    train_ds, validation_data=valid_ds, epochs=10)
```

Note Adaptive gradient descent algorithms including Adagrad, Adadelta, RMSprop, and Adam provide an alternative to the classical SGD algorithm. Adaptive algorithms provide a heuristic approach that automatically tunes hyperparameters, which avoids the expensive work inherent in tuning hyperparameters for a learning rate schedule manually.

Visualize:

```
visualize(range(10))
```

Generalize on Unseen Data

Generalize on the unseen test dataset for Xception:

```
x_model.evaluate(test_ds)
```

Generalize on the unseen test dataset for Inception:

```
i_model.evaluate(test_ds)
```

Stanford Dogs Experiment

We have demonstrated large dataset experiments, but none of them contained a lot of class labels. The *Stanford Dogs* dataset contains images of 120 breeds of dogs from around the world. The dataset is built with images and annotation from ImageNet for the task of fine-grained image categorization. Stanford Dogs contains 20,580 images split into 12,000 training images and 8,580 testing images. Class labels and bounding box annotations are provided for all 12,000 images.

Model Stanford Dogs with MobileNet

MobileNets are small, low-latency, low-power models parameterized to meet the resource constraints of a variety of use cases. Pre-trained models are considered low-power models because they consume low amounts of computer resources. The reason is that they have already been trained. Initial training does consume a lot of computer resources. But once trained, they consume a very low amount of computer resources. They can be built upon for classification, detection, embedding, and segmentation similar to the operations of other popular large-scale models like Inception.

The *MobileNet-v2* model was developed at Google. It is pre-trained on the ImageNet dataset, which is a large dataset consisting of 1.4 million images and 1,000 classes. *ImageNet* is a research training dataset with a wide variety of categories like jackfruit and syringe. Its base knowledge helps us classify dogs from our specific dataset.

Load Data

Load the training set from the Stanford Dogs training data (train split):

```
train_pups, dogs_info = tfds.load(
    'stanford_dogs', with_info=True,
    as_supervised=True, try_gcs=True,
    split='train')
```

Load validation and test sets with 50% splits each from the Stanford Dogs test data (test split):

```
(validation_pups, test_pups) = tfds.load(
    'stanford_dogs',
    split=['test[:50%]', 'test[50%:]'],
    as_supervised=True, try_gcs=True)
```

Metadata

Display metadata:

```
dogs_info
```

Visualize Examples

Create a function to get the *named* label from the integer label:

```
get_name = dogs_info.features['label'].int2str
```

By trial and error, we got all integer labels and converted them to named ones:

```
lbls = []
for image, label in train_pups.take(464):
  lbls.append(get_name(label))
set_lbl = set(lbls)
len(set_lbl)
```

Variable *set_lbl* holds the class labels for the dataset.

We know from the metadata that Stanford Dogs has 120 classes. So we adjusted (through trial and error) the number of examples taken until we got 120 unique labels!

Grab some images and labels for visualization:

```
img, lbl = [], []
for image, label in train_pups.take(9):
  img.append(image)
  lbl.append(get_name(label)[10:])
```

Display the first one:

```
lbl[0]
```

Display some examples as shown in Listing 7-5.

Listing 7-5. Display Examples

```
plt.figure(figsize=(12, 12))
for index in range(9):
  plt.subplot(3, 3, index + 1)
  plt.imshow(img[index])
  plt.title(lbl[index])
  plt.axis('off')
```

Display examples the easy way:

```
fig = tfds.show_examples(train_pups, dogs_info)
```

Check Image Shape

Take some examples and display image shape:

```
for img, lbl in train_pups.take(10):
  print (img.shape)
```

Since images vary in size, we must resize them:

Build the Input Pipeline

Get number of classes:

```
num_classes = dogs_info.features['label'].num_classes
num_classes
```

We already know the number of classes, but we show you how to get this number with one line of code.

Create pipeline variables:

```
IMG_LEN = 224
IMG_SHAPE = (IMG_LEN,IMG_LEN,3)
N_BREEDS = num_classes
```

Create a preprocessing function as shown in Listing 7-6.

Listing 7-6. Preprocessing Function

```
def preprocess(img, lbl):
  resized_image = tf.image.resize(img, [IMG_LEN, IMG_LEN])
  final_image = tf.keras.applications.mobilenet.preprocess_input(
      resized_image)
  label = tf.one_hot(lbl, N_BREEDS)
  return final_image, label
```

Create a function to build the input pipeline as shown in Listing 7-7.

Listing 7-7. Function to Build the Input Pipeline

```
def prepare(dataset, batch_size=None, shuffle_size=None):
  ds = dataset.map(preprocess, num_parallel_calls=4)
  ds = ds.shuffle(buffer_size=1000)
  if batch_size:
    ds = ds.batch(batch_size)
  if shuffle_size:
    ds = ds.shuffle(shuffle_size)
  ds = ds.prefetch(buffer_size=tf.data.experimental.AUTOTUNE)
  return ds
```

Build the pipeline:

```
BATCH_SIZE = 32
SHUFFLE_SIZE = 1000

train_dogs = prepare(train_pups, batch_size=BATCH_SIZE,
                     shuffle_size=SHUFFLE_SIZE)
validation_dogs = prepare(validation_pups, batch_size=32)
test_dogs = prepare(test_pups, batch_size=32)
```

Create the Model

Create the base model:

```
mobile_v2 = tf.keras.applications.MobileNetV2
mobile_model = mobile_v2(
    input_shape=IMG_SHAPE, include_top=False,
    weights='imagenet')
```

Explore base model layers:

```
tf.keras.utils.plot_model(
    mobile_model,
    show_shapes=True,
    show_layer_names=True)
```

Get number of layers:

```
len(mobile_model.layers)
```

Clear and seed:

```
tf.keras.backend.clear_session()
np.random.seed(0)
tf.random.set_seed(0)
```

Create the model (freeze the top layers):

```
mobile_model.trainable = False

sd_model = tf.keras.Sequential([
  mobile_model,
  GlobalAveragePooling2D(),
  Dropout(0.5),
  Dense(num_classes, activation='softmax')
])
```

Compile and Train

Compile and train the model as shown in Listing 7-8.

Listing 7-8. Compile and Train

```
EPOCHS = 5

sd_model.compile(
    optimizer=tf.keras.optimizers.Adamax(learning_rate=0.005),
    loss='categorical_crossentropy',
    metrics=['accuracy', 'top_k_categorical_accuracy'])

history = sd_model.fit(
    train_dogs, epochs=EPOCHS, validation_data=validation_dogs)
```

Since we are training many more images and many more classes, training time is much longer. So be patient. Don't be concerned if your computer crashes. It is not an error. More RAM is needed. We use Colab Pro and don't seem to have an issue.

Tip You may want to spend a few dollars a month to migrate to Colab Pro if you want to model real-world data.

Visualize:

```
visualize(range(EPOCHS))
```

Visualize the top five predictions as shown in Listing 7-9.

Listing 7-9. Top Five Predictions

```
acc = history.history['top_k_categorical_accuracy']
val_acc = history.history['val_top_k_categorical_accuracy']

epochs_range = range(EPOCHS)

plt.figure(figsize=(8, 8))
plt.subplot(1, 2, 1)
plt.plot(epochs_range, acc, label='Training Accuracy')
plt.plot(epochs_range, val_acc, label='Validation Accuracy')
plt.legend(loc='lower right')
plt.title('Top 5 Training and Validation Accuracy')
plt.grid(b=None)
```

Not bad! We get over 80% accuracy for breed detection. If we look at the top five predictions, the chance of guessing the correct breed jumps to over 97%! Since the dataset is large and complex, we show that using pre-trained models can have real-world use cases. Also it is amazing that we can build such a powerful model in just a few lines of code!

Model Trained Data with Unfrozen Layers

Unfreeze all the layers and continue training as shown in Listing 7-10.

Listing 7-10. Unfreeze All Layers and Continue Training

```
mobile_model.trainable = True

sd_model.compile(
    optimizer=tf.keras.optimizers.Adamax(0.00001),
    loss='categorical_crossentropy',
    metrics=['accuracy', 'top_k_categorical_accuracy'])

history = sd_model.fit(
    train_dogs, epochs=3,
    validation_data=validation_dogs)
```

We use a *much lower learning rate* to avoid damaging the pre-trained weights. We only run for three epochs because training time is so long.

Note Set learning rates much lower for real-world use cases like we did with Stanford Dogs.

Visualize:

```
visualize(range(3))
```

Generalize

Generalize from unseen data:

```
sd_model.evaluate(test_dogs)
```

Flowers Experiment

Let's work with a dataset stored as TFRecords. We strongly believe that practicing with a variety of datasets helps with learning.

Since we worked with flowers in earlier chapters, we won't describe the dataset in this section. We load flowers as TFRecords and use a pre-trained model for learning.

Read Flowers as TFRecords

Read TFRecord files from GCS:

```
piece1 = 'gs://flowers-public/'
piece2 = 'tfrecords-jpeg-192x192-2/*.tfrec'
TFR_GCS_PATTERN = piece1 + piece2
tfr_filenames = tf.io.gfile.glob(TFR_GCS_PATTERN)
```

Create Data Splits

Set pipeline and training parameters:

```
IMAGE_SIZE = [192, 192]
AUTO = tf.data.experimental.AUTOTUNE
BATCH_SIZE = 64
SHUFFLE_SIZE = 100
EPOCHS = 5
VALIDATION_SPLIT = 0.19
CLASSES = ['daisy', 'dandelion', 'roses', 'sunflowers', 'tulips']
```

Create splits as shown in Listing 7-11.

Listing 7-11. Create Splits

```
split = int(len(tfr_filenames) * VALIDATION_SPLIT)
training_filenames = tfr_filenames[split:]
validation_filenames = tfr_filenames[:split]
print ('Splitting dataset into {} training files and {}'
       'validation files'.\
       format(
           len(tfr_filenames), len(training_filenames),
           len(validation_filenames)), end = ' ')
print ('with a batch size of {}.'.format(BATCH_SIZE))

validation_steps = int(3670 // len(tfr_filenames) *\
                       len(validation_filenames)) // BATCH_SIZE
steps_per_epoch = int(3670 // len(tfr_filenames) *\
                      len(training_filenames)) // BATCH_SIZE
```

```
print ('There are {} batches per training epoch and {} '\
       'batches per validation run.'\
       .format(BATCH_SIZE, steps_per_epoch, validation_steps))
```

Create Functions to Load and Process TFRecord Files

Demonstrate one-hot encoding as shown in Listing 7-12.

Listing 7-12. Demonstrate One-Hot Encoding

```
named_lbl = 'sunflowers'
indx = CLASSES.index(named_lbl)
encode = tf.one_hot([indx], 5)
one_hot = encode[0].numpy()
print ('encoded label:', one_hot)
pos = tf.math.argmax(one_hot).numpy()
print ('integer label:', pos)
```

Use the tf.one_hot() API to encode a named label. Use the tf.math.argmax() API to convert the one-hot encoded label to an integer label.

Create a function to parse a TFRecord file as shown in Listing 7-13.

Listing 7-13. Function to Parse a TFRecord File

```
def read_tfrecord(example):
  features = {
      'image': tf.io.FixedLenFeature([], tf.string),
      'class': tf.io.FixedLenFeature([], tf.int64)
  }
  example = tf.io.parse_single_example(example, features)
  image = tf.image.decode_jpeg(example['image'], channels=3)
  image = tf.cast(image, tf.float32) / 255.0
  image = tf.reshape(image, [*IMAGE_SIZE, 3])
  class_label = example['class']
  one_hot = tf.one_hot(class_label, 5)
  return image, one_hot
```

Create a function to load TFRecord files as a tf.data.Dataset as shown in Listing 7-14.

Listing 7-14. Function to Load TFRecords as a tf.data.Dataset

```
def load_dataset(filenames):
  option_no_order = tf.data.Options()
  option_no_order.experimental_deterministic = False
  dataset = tf.data.TFRecordDataset(
      filenames, num_parallel_reads=AUTO)
  dataset = dataset.with_options(option_no_order)
  dataset = dataset.map(read_tfrecord, num_parallel_calls=AUTO)
  return dataset
```

Create a function to build an input pipeline from TFRecord files as shown in Listing 7-15.

Listing 7-15. Function to Build an Input Pipeline from TFRecord Files

```
def get_batched_dataset(filenames, train=False):
  dataset = load_dataset(filenames)
  dataset = dataset.cache()
  if train:
    dataset = dataset.repeat()
    dataset = dataset.shuffle(SHUFFLE_SIZE)
  dataset = dataset.batch(BATCH_SIZE)
  dataset = dataset.prefetch(AUTO)
  return dataset
```

Create Training and Test Sets

Instantiate the datasets:

```
training_dataset = get_batched_dataset(
    training_filenames, train=True)
validation_dataset = get_batched_dataset(
    validation_filenames, train=False)
training_dataset, validation_dataset
```

Display an image as shown in Listing 7-16.

Listing 7-16. Display an Image

```
for img, lbl in training_dataset.take(1):
  plt.axis('off')
  label = tf.math.argmax(lbl[0]).numpy()
  plt.title(CLASSES[label])
  fig = plt.imshow(img[0])
  tfr_flower_shape = img.shape[1:]
```

Create the Model

Create a list of pre-trained models:

```
ptm =\
  [tf.keras.applications.MobileNetV2,
   tf.keras.applications.VGG16,
   tf.keras.applications.MobileNet,
   tf.keras.applications.xception.Xception,
   tf.keras.applications.InceptionV3,
   tf.keras.applications.ResNet50]
```

Choose any of the pre-trained models by index. We use Xception to create a base model:

```
pre_trained_model = ptm[3](
    weights='imagenet', include_top=False,
    input_shape=[*IMAGE_SIZE, 3])
```

Clear and seed:

```
tf.keras.backend.clear_session()
np.random.seed(0)
tf.random.set_seed(0)
```

Create the final model:

```
pre_trained_model.trainable = True

flower_model = tf.keras.Sequential([
  pre_trained_model,
```

```
    GlobalAveragePooling2D(),
    Dense(5, activation='softmax')])
```

We use the Xception pre-trained model. We drop the ImageNet-specific top layers with *include_top=false* and a max pooling and a softmax layer to predict the five flower classes. Notice that we also unfreeze all of the top layers! Don't try unfreezing all layers up front with larger datasets!

Compile and Train

Compile:

```
optimizer = tf.keras.optimizers.Adam(learning_rate=0.0001)
```

```
flower_model.compile(
    optimizer=optimizer,
    loss = 'categorical_crossentropy',
    metrics=['accuracy'])
```

The initial learning rate is often the single most important hyperparameter. If one can tune only one hyperparameter, the learning rate is the one worth tuning. Conveniently, we can use the *Adam* optimizer to automatically tune the learning rate! But we still must set the initial learning rate.

We set a very low learning rate so that the model (hopefully) learns a more optimal or even a globally optimal set of weights. But training takes significantly longer. We set the learning rate to a low value to ensure that gradient descent does not increase training error. Initially, we want to allow the neural network to randomly adjust its weights. Lower learning rates increase randomization. Higher learning rates decrease randomization.

Train:

```
history = flower_model.fit(
    training_dataset, epochs=EPOCHS,
    verbose=1, steps_per_epoch=steps_per_epoch,
    validation_steps=validation_steps,
    validation_data=validation_dataset)
```

Visualize

Visualize training performance:

```
visualize(range(EPOCHS))
```

Generalize

We generalize on the validation set because we didn't split out a test one:

```
flower_model.evaluate(validation_dataset)
```

Rock-Paper-Scissors Experiment

The final experiment is with the *rock_paper_scissors* dataset. The dataset is adapted from the rock-paper-scissors game. The data contains images of hands playing the rock-paper-scissors game.

Rock-paper-scissors is a hand game usually played between two people where each player simultaneously forms one of three shapes with an outstretched hand. The possible shapes are rock, paper, and scissors. The rules are simple. Rock beats scissors. Paper beats rock. Scissors beat paper. Metaphorically rock smashes scissors, paper covers rock, and scissors cut paper.

Load the Data

Load the training set:

```
train_digits, rps_info = tfds.load(
    'rock_paper_scissors', with_info=True,
    split='train', as_supervised=True,
    try_gcs=True)
```

Load the test set:

```
test_digits = tfds.load(
    'rock_paper_scissors',  try_gcs=True,
    as_supervised=True, split='test')
```

Inspect tensors:

```
for image, label in train_digits.take(5):
  print (image.shape, label.numpy())
```

Display metadata:

```
rps_info
```

Visualize

Display some examples:

```
fig = tfds.show_examples(train_digits, rps_info)
```

Build the Input Pipeline

Create a function to process images and labels as shown in Listing 7-17.

Listing 7-17. Function to Process Data

```
def process_digits(image, label):
  resized_image = tf.image.resize(image, [224, 224])
  final_image = tf.keras.applications.xception.\
                preprocess_input(resized_image)
  one_hot = tf.one_hot(label, 3)
  return final_image, one_hot
```

Labels are expected as one-hot encoded objects.

Build the pipeline:

```
BATCH_SIZE = 64
shuffle = 250

train_fingers = train_digits.shuffle(shuffle).\
  map(process_digits).batch(BATCH_SIZE).prefetch(1)
test_fingers = test_digits.map(process_digits).\
  batch(BATCH_SIZE).prefetch(1)
```

Create the Model

Create the base model:

```
Xception = tf.keras.applications.xception.Xception
xception_model = Xception(
    weights='imagenet', include_top=False)
```

Clear and seed:

```
tf.keras.backend.clear_session()
np.random.seed(0)
tf.random.set_seed(0)
```

Create the final model:

```
pre_trained_model.trainable = True

fingers_model = tf.keras.Sequential([
  xception_model,
  GlobalAveragePooling2D(),
  Dense(3, activation='softmax')])
```

Unfreeze all layers!

Compile and Train

Compile:

```
optimizer = tf.keras.optimizers.Adam(learning_rate=0.00001)

fingers_model.compile(
    optimizer=optimizer,
    loss = 'categorical_crossentropy',
    metrics=['accuracy'])
```

Since we unfreeze all layers, set a REALLY low learning rate to allow the network to randomly balance neuron weights during early training epochs.

Train:

```
history = fingers_model.fit(
    train_fingers, epochs=10,
    validation_data=test_fingers)
```

Not bad!

Visualize

Visualize training performance:

```
visualize(range(10))
```

Generalize

Generalize on test data:

```
fingers_model.evaluate(test_fingers)
```

Tips and Concepts

For additional tips to tune transfer learning models, peruse

https://medium.com/@kenneth.ca95/a-guide-to-transfer-learning-with-keras-using-resnet50-a81a4a28084b

For a comprehensive (yet readable) take on the subject, peruse

https://towardsdatascience.com/a-comprehensive-hands-on-guide-to-transfer-learning-with-real-world-applications-in-deep-learning-212bf3b2f27a

CHAPTER 8

Stacked Autoencoders

The first seven chapters focused on supervised learning algorithms. **Supervised learning** is a subcategory of ML that uses *labeled* datasets to train algorithms to classify data and predict outcomes accurately. The remaining chapters focus on unsupervised learning algorithms. **Unsupervised learning** uses ML algorithms to analyze and cluster *unlabeled* datasets. Such algorithms discover hidden patterns or data groupings *without the need for human intervention*.

Autoencoders are artificial neural networks that learn dense representations of input data without any supervision. Learned dense representations are often called latent representations (or codings). Codings are used to reconstruct the original output.

Codings typically have much lower dimensionality than the input data, which makes autoencoders useful for dimensionality reduction. Autoencoders can also be used for feature extraction, unsupervised pre-training of deep neural networks, and generative models. As generative models, they can randomly generate new data that looks very similar to the training data.

It may seem counterintuitive to train autoencoders if they just produce the same output from input data. But we show how a trained autoencoder can reconstruct images from *new* data! We also show how they are effective dimensionality reduction mechanisms.

An autoencoder consists of an encoder component, a code component, and a decoder component. The encoder compresses the input and produces the code. The decoder then reconstructs the original input from the code.

Notebooks for chapters are located at the following URL:

`https://github.com/paperd/deep-learning-models`

We present four stacked autoencoder experiments. We begin with a basic stacked encoder experiment. We continue with a tying weights experiment. We end with denoising and simple tuning experiments.

D. Paper, *State-of-the-Art Deep Learning Models in TensorFlow*, https://doi.org/10.1007/978-1-4842-7341-8_8

We start with a simple stacked encoder because it is the most basic type of autoencoder. Each subsequent experiment adds complexity demonstrated with tying weights, denoising, and tuning autoencoders.

Begin setting up the Colab ecosystem by importing the main TensorFlow library and instantiating the GPU.

Import the TensorFlow Library

Import the library and alias it as **tf**:

```
import tensorflow as tf
```

Aliasing the TensorFlow library as tf is common practice.

GPU Hardware Accelerator

As a convenience, we include the steps to enable the GPU in a Colab notebook:

1. Click *Runtime* in the top-left menu.
2. Click *Change runtime type* from the drop-down menu.
3. Choose *GPU* from the *Hardware accelerator* drop-down menu.
4. Click *Save*.

Verify that the GPU is active:

```
tf.__version__, tf.test.gpu_device_name()
```

If '/device:GPU:0' is displayed, the GPU is active. If '..' is displayed, the regular CPU is active.

Note If you get the error **NAME 'TF' IS NOT DEFINED**, re-execute the code to import the TensorFlow library!

Basic Stacked Encoder Experiment

Stacked encoders have multiple hidden layers. The architecture is typically symmetrical with regard to the central hidden layer, which is called the *coding layer*. We show you how to create and train a stacked encoder in this experiment.

Load Data

Load Fashion-MNIST as NumPy arrays:

```
import tensorflow_datasets as tfds

(x_train_img, _), (x_test_img, _) = tfds.as_numpy(
    tfds.load('fashion_mnist', split=['train','test'],
              batch_size=-1, as_supervised=True,
              try_gcs=True))
```

Notice that we don't load the class labels from Fashion-MNIST because autoencoders are unsupervised models! That is, they *don't need labeled data to learn!*

Scale Data

Scale by dividing datasets by the number of pixels that represent an image:

```
import numpy as np

x_train, x_test = x_train_img.astype(np.float32) / 255.,\
                  x_test_img.astype(np.float32) / 255.
```

Scaling data is recommended for deep learning experiments because it typically smooths optimization during training.

Build the Stacked Encoder Model

Clear and seed:

```
tf.keras.backend.clear_session()
np.random.seed(0)
tf.random.set_seed(0)
```

Get input shape:

```
in_shape = x_train.shape[1:]
in_shape
```

Import libraries:

```
from tensorflow.keras.models import Sequential
from tensorflow.keras.layers import Dense, Flatten,\
  Reshape
```

An autoencoder is split into an encoder and a decoder. The encoder encodes examples into smaller denser representations. The decoder takes these dense representations and tries to reconstruct the output back to the original input.

Create the encoder:

```
stacked_encoder = Sequential([
  Flatten(input_shape=in_shape),
  Dense(128, activation='relu'),
  Dense(64, activation='relu'),
  Dense(32, activation='relu')
])
```

The encoder accepts 28 × 28-pixel grayscale images, flattens them so that each image is represented as a vector of size 784, and processes the vectors through three Dense layers of diminishing sizes (128 units to 64 units to 32 units). The 32-unit layer is the coding layer (central hidden layer). For each input image, the encoder outputs a vector of size 32.

Create the decoder:

```
stacked_decoder = Sequential([
  Dense(64, activation='relu'),
  Dense(128, activation='relu'),
  Dense(28 * 28, activation='sigmoid'),
  Reshape(in_shape)
])
```

The decoder accepts codings of size 32 from the encoder and processes them through three Dense layers of increasing sizes (64 units to 128 units to 784 units). It then reshapes the final vectors into 28 × 28 arrays so the decoder's outputs have the same shape as the encoder's inputs.

Create the stacked autoencoder:

```
stacked_ae = Sequential([stacked_encoder, stacked_decoder])
```

Compile and Train

Create a function for the accuracy metric:

```
def rounded_accuracy(y_true, y_pred):
    return tf.keras.metrics.binary_accuracy(tf.round(y_true),
                                            tf.round(y_pred))
```

Reconstruction is a binary problem. Either outputs match inputs or they do not match. Reconstruction loss penalizes a network for creating outputs different from inputs. So binary accuracy is ideal for this experiment.

Compile:

```
opt = tf.keras.optimizers.SGD(lr=1.5)

stacked_ae.compile(
    loss='binary_crossentropy',
    optimizer=opt, metrics=[rounded_accuracy])
```

Train:

```
sae_history = stacked_ae.fit(
    x_train, x_train, epochs=10,
    validation_data=(x_test, x_test))
```

Visualize Performance

Import a plotting library:

```
import matplotlib.pyplot as plt
```

Create a plotting function as shown in Listing 8-1.

Listing 8-1. Training Performance Visualization Function

```
def viz_history(training_history):
  loss = training_history.history['loss']
  val_loss = training_history.history['val_loss']
  accuracy = training_history.history['rounded_accuracy']
  val_accuracy = training_history.history['val_rounded_accuracy']
  plt.figure(figsize=(14, 4))
  plt.subplot(1, 2, 1)
  plt.title('Loss')
  plt.xlabel('Epoch')
  plt.ylabel('Loss')
  plt.plot(loss, label='Training set')
  plt.plot(val_loss, label='Test set', linestyle='--')
  plt.legend()
  plt.grid(linestyle='--', linewidth=1, alpha=0.5)
  plt.subplot(1, 2, 2)
  plt.title('Accuracy')
  plt.xlabel('Epoch')
  plt.ylabel('Accuracy')
  plt.plot(accuracy, label='Training set')
  plt.plot(val_accuracy, label='Test set', linestyle='--')
  plt.legend()
  plt.grid(linestyle='--', linewidth=1, alpha=0.5)
  plt.show()
```

Visualize:

```
viz_history(sae_history)
```

Visualize the Reconstructions

Create a function to plot a grayscale 28 × 28 image:

```
def plot_image(image):
    plt.imshow(image, cmap='binary')
    plt.axis('off')
```

Create a function to visualize original images and reconstructions as shown in Listing 8-2.

Listing 8-2. Visualization Function for Reconstructions

```
def show_reconstructions(model, images, n_images):
  reconstructions = model.predict(images[:n_images])
  reconstructions = tf.squeeze(reconstructions)
  fig = plt.figure(figsize=(n_images * 1.5, 3))
  for image_index in range(n_images):
    plt.subplot(2, n_images, 1 + image_index)
    plot_image(images[image_index])
    plt.subplot(2, n_images, 1 + n_images + image_index)
    plot_image(reconstructions[image_index])
```

The function accepts the trained model, a set of *test* images, and a number that represents a batch of images. The function computes predictions for the batch of *test* images. It then squeezes out the *1* dimension from the predictions for visualization purposes. Keep in mind that our autoencoder model is trained on the training data (*x_train*) and reconstructions are based on the test data (*x_test*) from the Fashion-MNIST dataset. Also be aware that reconstructions are created *without any labeled data.*

Check the dimensionality of test data:

```
x_test.shape
```

Shape is (10000, 28, 28, 1). So there are 10,000 28 × 28 × 1 images in the test dataset.

To visualize with the imshow() function, remove dimensions of size "1" from the tensor:

```
x_test_imgs = tf.squeeze(x_test)
x_test_imgs.shape
```

The shape is now TensorShape([10000, 28, 28]), which is what the imshow() function expects for images.

Visualize:

```
show_reconstructions(stacked_ae, x_test_imgs, 6)
```

Reconstructed images are generated from test images based on predictions from the trained model. The autoencoder is able to learn how to decompose images into small bits of data. The small bits of data provide representations of the images. The autoencoder learns how to reconstruct original images from these representations. Of course, the reconstructions are not exactly the same as the originals because we use a simple stacked autoencoder.

Breakdown

Let's break down the function in Listing 8-2 to see how it works.

Grab the first image from the test set as a batch of one:

```
img = x_test[:1]
```

Since prediction method computations are done in batches, grab the first image as a batch of one.

Make a prediction based on the first image batch:

```
reconstruction = stacked_ae.predict(img)
```

Drop the "1" dimension:

```
reconstruction = tf.squeeze(reconstruction)
```

Plot the reconstruction:

```
plot_image(reconstruction)
```

Plot the actual image:

```
plot_image(tf.squeeze(x_test[0]))
```

Squeeze out the "1" dimension from the image to plot.

Dimensionality Reduction

Autoencoders are useful for dimensionality reduction. **Dimensionality reduction** is the process of reducing the number of random variables under consideration (or input features) in a deep learning experiment by obtaining a set of principal variables. Principal variables are the most important features in a dataset. Dimensionality reduction is commonly used in deep learning because large numbers of input features can cause poor performance for ML algorithms.

To perform dimensionality reduction, *we need labels.* So load labels from the test dataset:

```
test = tfds.as_numpy(
    tfds.load('fashion_mnist', split=['test'],
              batch_size=-1, as_supervised=True,
              try_gcs=True))
```

Slice test labels from the test dataset:

```
y_test = test[0][1]
```

Use the encoder to reduce dimensionality as shown in Listing 8-3.

Listing 8-3. Dimensionality Reduction Algorithm

```
from sklearn.manifold import TSNE

np.random.seed(0)
x_test_compressed = stacked_encoder.predict(x_test_imgs)
tsne = TSNE()
x_test_2D = tsne.fit_transform(x_test_compressed)
x_test_2D = (x_test_2D - x_test_2D.min()) /\
  (x_test_2D.max() - x_test_2D.min())
```

Use scikit-learn's implementation of the t-SNE (t-distributed stochastic neighbor embedding) algorithm to reduce dimensionality to 2D for visualization. The algorithm reduces dimensionality to 32. The *t-SNE* algorithm is a statistical method for visualizing high-dimensional data by giving each data point a location in a two- or three-dimensional map.

Visualize:

```
plt.scatter(x_test_2D[:, 0], x_test_2D[:, 1],
            c=y_test, s=10, cmap='tab10')
plt.axis('off')
plt.show()
```

Each class is represented by a different color.

Create a prettier visualization as shown in Listing 8-4.

Listing 8-4. Pretty Dimensionality Reduction Visualization

```
import matplotlib as mpl

plt.figure(figsize=(10, 8))
cmap = plt.cm.tab10
plt.scatter(x_test_2D[:, 0], x_test_2D[:, 1],
            c=y_test, s=10, cmap=cmap)
```

```
image_positions = np.array([[1., 1.]])
for index, position in enumerate(x_test_2D):
    dist = np.sum((position - image_positions) ** 2, axis=1)
    if np.min(dist) > 0.02: # if far enough from other images
        image_positions = np.r_[image_positions, [position]]
        imagebox = mpl.offsetbox.AnnotationBbox(
            mpl.offsetbox.OffsetImage(x_test_imgs[index],
                                      cmap='binary'),
            position, bboxprops={
                'edgecolor': cmap(y_test[index]), 'lw': 2})
        plt.gca().add_artist(imagebox)
plt.axis('off')
plt.show()
```

With this visualization, we can see extracted class labels with the help of dimensionality reduction.

Tying Weights Experiment

When an autoencoder is neatly symmetrical, we can tie the weights of the decoder layers to the weights of the encoder layers. *Neatly symmetrical* means that the encoder and decoder have symmetrical layer construction. So we halve the number of weights in the model, which speeds training and reduces overfitting.

An autoencoder with tied weights has decoder weights that are the transpose of the encoder weights, which is a form of parameter sharing. We reduce the number of parameters with parameter sharing.

Define a Custom Layer

To tie the weights of the encoder and the decoder, use the transpose of the encoder's weights as the decoder weights as shown in Listing 8-5.

Listing 8-5. Custom Layer Class

```
class DenseTranspose(tf.keras.layers.Layer):
  def __init__(self, dense, activation=None, **kwargs):
    self.dense = dense
    self.activation = tf.keras.activations.get(activation)
    super().__init__(**kwargs)
  def build(self, batch_input_shape):
    self.biases = self.add_weight(
        name='bias', shape=[self.dense.input_shape[-1]],
        initializer='zeros')
    super().build(batch_input_shape)
  def call(self, inputs):
    z = tf.matmul(
        inputs, self.dense.weights[0], transpose_b=True)
    return self.activation(z + self.biases)
```

The class accepts a layer from a model and an activation function (if included in a layer) and transposes the data. A lot of times we have to preprocess data fed into ML algorithms. The reason is that data may be stored as rows, but the ML algorithm expects input as columns or vice versa. So transposition is a very useful operation in ML activities.

Build the Tied Weights Model

Clear and seed:

```
tf.keras.backend.clear_session()
np.random.seed(0)
tf.random.set_seed(0)
```

For convenience, create three Dense layers for the model:

```
dense_1 = Dense(128, activation='relu')
dense_2 = Dense(64, activation='relu')
dense_3 = Dense(32, activation='relu')
```

We optionally create these variables to use in the encoder. Of course, you can include the layer syntax directly in the model if you wish.

Build the encoder:

```
tied_encoder = Sequential([
  Flatten(input_shape=in_shape),
  dense_1,
  dense_2,
  dense_3
])
```

Build the decoder:

```
tied_decoder = Sequential([
  DenseTranspose(dense_3, activation='relu'),
  DenseTranspose(dense_2, activation='relu'),
  DenseTranspose(dense_1, activation='sigmoid'),
  Reshape([28, 28])
])
```

Since the autoencoder is neatly symmetrical, we can map the DenseTranspose class onto the layers from the encoder. A DenseTranspose layer is similar to a Dense one, but the input matrix is stored in column-major order rather than row-major order, That is, it is the transpose of an ordinary Dense matrix.

Build the tied weights autoencoder:

```
tied_ae = Sequential([tied_encoder, tied_decoder])
```

Compile and Train

Compile:

```
tied_ae.compile(loss='binary_crossentropy',
                optimizer=opt, metrics=[rounded_accuracy])
```

Train:

```
tied_history = tied_ae.fit(
    x_train, x_train, epochs=10,
    validation_data=(x_test, x_test))
```

Visualize Training Performance

Visualize:

```
viz_history(tied_history)
```

Visualize Reconstructions

Show test image reconstructions based on predictions from the trained model:

```
show_reconstructions(tied_ae, x_test_imgs, 6)
plt.show()
```

Denoising Experiment

An autoencoder can also be trained to remove noise from images. The idea is to add *random noise* to inputs and train to recover the original noise-free inputs. This sounds counterintuitive, but it works!

Noise inherent in images can create trouble because algorithms may believe that the noise is a pattern that should be learned. But adding random noise keeps the network from memorizing training samples because they are changing all of the time.

Random noise allows us to test the robustness and performance of an algorithm in the presence of known amounts of noise. That is, we know what noise we are adding. By adding noise to input data, we get more data for which our deep neural network can train. But training on noisy data means that the model generalizes on noisy data.

Build the Denoising Model

Clear and seed:

```
tf.keras.backend.clear_session()
np.random.seed(0)
tf.random.set_seed(0)
```

Import the library for Gaussian noise:

```
from tensorflow.keras.layers import GaussianNoise
```

The library enables us to apply additive zero-centered Gaussian noise. **Gaussian noise** (named after Carl Friedrich Gauss) is statistical noise with a probability density function equal to that of the normal distribution (also known as the Gaussian distribution). So the values of added noise are Gaussian-distributed (or normally distributed).

We add Gaussian noise because it is the basic noise model used in information theory to mimic the effect of many random processes that occur naturally in nature. And Gaussian noise is normally distributed noise, so it fits many natural phenomena.

Add pure Gaussian noise directly to the encoder:

```
gaussian_encoder = Sequential([
  Flatten(input_shape=in_shape),
  GaussianNoise(0.2),
  dense_1,
  dense_2,
  dense_3
])
```

Build the decoder:

```
gaussian_decoder = Sequential([
  DenseTranspose(dense_3, activation='relu'),
  DenseTranspose(dense_2, activation='relu'),
  DenseTranspose(dense_1, activation='sigmoid'),
  Reshape([28, 28])
])
```

For better performance, tie the weights of the decoder layers to the weights of the encoder layers.

Build the denoising autoencoder:

```
gaussian_ae = Sequential([gaussian_encoder, gaussian_decoder])
```

Compile and Train

Compile:

```
gaussian_ae.compile(
    loss='binary_crossentropy',
    optimizer=opt, metrics=[rounded_accuracy])
```

Train:

```
gae_history = gaussian_ae.fit(
    x_train, x_train, epochs=10,
    validation_data=(x_test, x_test))
```

Visualize Training Performance

Visualize:

```
viz_history(gae_history)
```

Visualize Reconstructions

Add the same amount of Gaussian noise to test images:

```
noise = GaussianNoise(0.2)
show_reconstructions(gaussian_ae, noise(x_test_imgs), 6)
plt.show()
```

Dropout Experiment

Dropout is a regularization technique that helps prevent overfitting. Dropout randomly *turns off* some neurons during training to turn a neural network into an ensemble of neural networks. The idea is to approximate training a large number of neural networks with different architectures in parallel.

The effect of Dropout makes the layer look like and be treated like a layer with a different number of nodes and connectivity to the prior layer. So each update to a layer during training is performed with a *different view* of the configured layer.

However, any amount of Dropout represents an information loss! By setting a layer to a Dropout probability of 0.5, we lose half of the information at that layer during each epoch (or training iteration)! Using Dropout is recommended in most cases, but not at every layer, and it should not be set above 0.5.

Dropout makes the training process noisy, which forces nodes within a layer to probabilistically take on more or less responsibility for the inputs. By dropping out a neuron, it is temporarily removed from the network along with all of its incoming and outgoing connections.

Dropout is implemented per layer in a neural network. It can be used with most types of layers such as Dense fully connected layers, convolutional layers, and recurrent layers.

Dropout may be implemented on any or all hidden layers. A hyperparameter specifies the probability that outputs of the layer are dropped out. A common value for hidden layers is a probability of 0.5, which seems to be close to optimal for a wide range of networks and tasks. For input layers, the optimal value must be closer to 0 than to 0.5.

The experiment demonstrates how easy it is to build and train a Dropout autoencoder. The only change is to add a Dropout value in the encoder.

Build the Dropout Model

Clear and seed:

```
tf.keras.backend.clear_session()
np.random.seed(0)
tf.random.set_seed(0)
```

Import the library for Dropout:

```
from tensorflow.keras.layers import Dropout
```

Build the encoder with Dropout of 0.5:

```
dropout_encoder = Sequential([
  Flatten(input_shape=in_shape),
  Dropout(0.5),
  dense_1,
  dense_2,
  dense_3
])
```

Build the decoder with tied weights for better performance:

```
dropout_decoder = Sequential([
  DenseTranspose(dense_3, activation='relu'),
  DenseTranspose(dense_2, activation='relu'),
```

```
  DenseTranspose(dense_1, activation='sigmoid'),
  Reshape([28, 28])
])
```

Build the Dropout autoencoder:

```
dropout_ae = Sequential([dropout_encoder, dropout_decoder])
```

Compile and Train

Compile:

```
dropout_ae.compile(
    loss='binary_crossentropy',
    optimizer=opt, metrics=[rounded_accuracy])
```

Train:

```
drop_history = dropout_ae.fit(
    x_train, x_train, epochs=10,
    validation_data=(x_test, x_test))
```

Visualize Training Performance

Visualize:

```
viz_history(drop_history)
```

Visualize Reconstructions

Add the same amount of Dropout noise to test images:

```
dropout = Dropout(0.5)
show_reconstructions(dropout_ae, dropout(x_test_imgs), 6)
plt.show()
```

Summary

In this chapter, we demonstrate basic stacked autoencoders with a few tweaks. In the next chapter, we introduce more powerful autoencoders.

CHAPTER 9

Convolutional and Variational Autoencoders

Autoencoders don't typically work well with images unless they are very small. But convolutional and variational autoencoders works much better than a feedforward dense ones with large color images.

A variational autoencoder (VAE) is a deep learning technique for learning latent representations. They apply learned latent space representations to draw images and interpolate between sentences. A VAE works by compressing the input into a latent space representation and then reconstructing the output from this representation. **Latent space** is a mathematical representation of compressed data.

A VAE tackles the problem of latent space irregularity by making the encoder return a Gaussian distribution over the latent space instead of a single point and adding regularization in the loss function over that returned distribution. In deep learning, data can be compressed to better manage computational resources. But the latent space must be spread over a Gaussian distribution to normalize it for model consumption.

A major advantage of a VAE is that it describes an observation (in latent space) in a Gaussian probabilistic manner to better simulate a natural phenomenon. So rather than building an encoder that outputs a single value to describe each latent state attribute, a VAE builds one to describe a probability distribution for each latent attribute. The result is a generated image of much higher quality.

Notebooks for chapters are located at the following URL:

https://github.com/paperd/deep-learning-models

We demonstrate convolutional and variational autoencoders with code examples. We begin with a convolutional autoencoder experiment. We continue with a VAE experiment. We end with a VAE with Tensorflow Probability (TFP) Layers experiment. The three experiments show their superiority over basic stacked autoencoders by producing much clearer renditions of the original images.

D. Paper, *State-of-the-Art Deep Learning Models in TensorFlow*, https://doi.org/10.1007/978-1-4842-7341-8_9

Begin setting up the Colab ecosystem by importing the main TensorFlow library and instantiating the GPU.

Import the TensorFlow Library

Import the library and alias it as **tf**:

```
import tensorflow as tf
```

Aliasing the TensorFlow library as tf is common practice.

GPU Hardware Accelerator

For convenience, we include the steps to enable the GPU in a Colab notebook:

1. Click *Runtime* in the top-left menu.
2. Click *Change runtime type* from the drop-down menu.
3. Choose *GPU* from the *Hardware accelerator* drop-down menu.
4. Click *Save.*

Verify that the GPU is active:

```
tf.__version__, tf.test.gpu_device_name()
```

If '/device:GPU:0' is displayed, the GPU is active. If '..' is displayed, the regular CPU is active.

Note If you get the error **NAME 'TF' IS NOT DEFINED**, re-execute the code to import the TensorFlow library!

Convolutional Encoder Experiment

A *convolutional autoencoder* is a variant of a convolutional neural network used for unsupervised learning of convolutional filters. Convolutional autoencoders minimize reconstruction errors by learning the optimal filters during image reconstruction. In the experiment, the convolutional autoencoder learns how to generate new images as facsimiles of the input features from a dataset.

Load Data

We use the *horses_or_humans* dataset for this experiment. The horses_or_humans dataset contains 1,283 horse and human images.

Load the dataset as a TFDS object for inspection:

```
import tensorflow_datasets as tfds

data, hh_info = tfds.load(
    'horses_or_humans', with_info=True,
    split='train', try_gcs=True)
```

Inspect Data

Get metadata:

```
hh_info
```

Get labels and number of classes:

```
class_labels = hh_info.features['label'].names
num_classes = hh_info.features['label'].num_classes
class_labels, num_classes
```

Display Examples

Show some examples with the tfds API:

```
fig = tfds.show_examples(data, hh_info)
```

Display examples as a pandas dataframe:

```
tfds.as_dataframe(data.take(4), hh_info)
```

Display an image manually:

```
import matplotlib.pyplot as plt

for element in data.take(1):
  plt.imshow(element['image'])
  plt.axis('off')
```

Display a grid of examples manually as shown in Listing 9-1.

Listing 9-1. Grid of Examples

```
img, lbl = [], []
for element in data.take(9):
  img.append(element['image'])
  lbl.append(element['label'].numpy())
fig=plt.figure(figsize=(8, 8))
columns = 3
rows = 3
for i in range(1, columns*rows+1):
  fig.add_subplot(rows, columns, i)
  plt.imshow(img[i-1])
  plt.title(class_labels[lbl[i-1]])
  plt.axis('off')
plt.show()
```

Create a set of images and labels. Plot the images and labels in a grid.

Get Training Data

From the metadata, we know the data splits:

```
(x_train_img, _), (x_test_img, _) = tfds.as_numpy(
    tfds.load(
        'horses_or_humans', split=['train','test'],
        batch_size=-1, as_supervised=True,
        try_gcs=True))
```

Since autoencoders are unsupervised models, we don't need the labels.

Get number of training and test elements:

```
len(x_train_img), len(x_test_img)
```

Inspect Shapes

Inspect train and test shapes:

```
x_train_img.shape, x_test_img.shape
```

Inspect image size:

```
for element in range(10):
  print (x_train_img.shape)
```

Since images are the same shape, we don't have to resize them for training.

Preprocess Image Data

Scale images:

```
import numpy as np

x_train, x_test = x_train_img.astype(np.float32) / 255,\
                  x_test_img.astype(np.float32) / 255
```

Inspect a vector from a training image before and after scaling:

```
x_train_img[0][0][0], x_train[0][0][0]
```

Create a Convolutional Autoencoder

Clear and seed:

```
tf.keras.backend.clear_session()
np.random.seed(0)
tf.random.set_seed(0)
```

Get input shape:

```
hh_shape = hh_info.features['image'].shape
hh_shape
```

Import libraries:

```
from tensorflow.keras.models import Sequential
from tensorflow.keras.layers import Conv2D, MaxPool2D,\
  Dense, Flatten, Input, Conv2DTranspose, Reshape
from tensorflow.keras.models import Model
```

Create an encoder as shown in Listing 9-2.

Listing 9-2. Encoder for a Convolutional Autoencoder

```
conv_encoder = Sequential([
  Input(shape=hh_shape),
  Conv2D(16, kernel_size=3, padding='SAME', activation='selu'),
  MaxPool2D(pool_size=2),
  Conv2D(32, kernel_size=3, padding='SAME', activation='selu'),
  MaxPool2D(pool_size=2),
  Conv2D(64, kernel_size=3, padding='SAME', activation='selu'),
  MaxPool2D(pool_size=2)
])
```

The encoder is composed of convolutional layers and pooling layers. The encoder reduces the spatial dimensionality of the inputs (height and width) while increasing the depth (number of feature maps).

Create the decoder as shown in Listing 9-3.

Listing 9-3. Decoder for a Convolutional Autoencoder

```
conv_decoder = Sequential([
  Conv2DTranspose(32, kernel_size=3, strides=2, padding='VALID',
                  activation='selu'),
  Conv2DTranspose(16, kernel_size=3, strides=2, padding='SAME',
                  activation='selu'),
```

```
  Conv2DTranspose(3, kernel_size=3, strides=2, padding='SAME',
                  activation='sigmoid')
])
```

The decoder does the reverse of the encoder by upscaling images and reducing their depth back to their original dimensions. Conv2DTranspose layers are used for this purpose.

Create the convolutional autoencoder:

```
conv_ae = Sequential([conv_encoder, conv_decoder])
```

Compile and Train

Create a function for the accuracy metric:

```
def rounded_accuracy(y_true, y_pred):
    return tf.keras.metrics.binary_accuracy(
        tf.round(y_true), tf.round(y_pred))
```

Compile:

```
conv_ae.compile(
    loss='binary_crossentropy',
    optimizer=tf.keras.optimizers.SGD(lr=1.0),
    metrics=[rounded_accuracy])
```

Train:

```
cae_history = conv_ae.fit(
    x_train, x_train, epochs=5,
    validation_data=(x_test, x_test))
```

Visualize Training Performance

Create a visualization function as shown in Listing 9-4.

Listing 9-4. Performance Visualization Function

```
def viz_history(training_history):
  loss = training_history.history['loss']
  val_loss = training_history.history['val_loss']
  accuracy = training_history.history['rounded_accuracy']
  val_accuracy = training_history.history['val_rounded_accuracy']
  plt.figure(figsize=(14, 4))
  plt.subplot(1, 2, 1)
  plt.title('Loss')
  plt.xlabel('Epoch')
  plt.ylabel('Loss')
  plt.plot(loss, label='Training set')
  plt.plot(val_loss, label='Test set', linestyle='--')
  plt.legend()
  plt.grid(linestyle='--', linewidth=1, alpha=0.5)
  plt.subplot(1, 2, 2)
  plt.title('Accuracy')
  plt.xlabel('Epoch')
  plt.ylabel('Accuracy')
  plt.plot(accuracy, label='Training set')
  plt.plot(val_accuracy, label='Test set', linestyle='--')
  plt.legend()
  plt.grid(linestyle='--', linewidth=1, alpha=0.5)
  plt.show()
```

Invoke the visualization function:

```
viz_history(cae_history)
```

Visualize Reconstructions

Create a reconstruction visualization function as shown in Listing 9-5.

Listing 9-5. Reconstruction Visualization Function

```
def show_reconstructions(model, images, n_images, reshape=False):
  reconstructions = model.predict(images[:n_images])
  if reshape:
    reconstructions = tf.squeeze(reconstructions)
  fig = plt.figure(figsize=(n_images * 1.5, 3))
  for image_index in range(n_images):
    plt.subplot(2, n_images, 1 + image_index)
    plot_image(images[image_index])
    plt.subplot(2, n_images, 1 + n_images + image_index)
    plot_image(reconstructions[image_index])
```

The function generates reconstructions from image predictions.

Create a function to show an image:

```
def plot_image(image):
  plt.imshow(image, cmap='binary')
  plt.axis('off')
```

Show original and reconstructed images:

```
show_reconstructions(conv_ae, x_test, 5)
```

Variational Autoencoder Experiment

A **variational autoencoder** (VAE) is an unsupervised ML technique that probabilistically learns efficient data codings. VAE training is regularized to mitigate overfitting and ensure that the latent space effectively generates similar outputs when compared to training inputs. A VAE is very different from other autoencoders because it encodes input as a distribution over the latent space rather as a single point.

Initially, input is encoded as a distribution over the latent space. A point from the latent space is then sampled from the distribution. The sampled point is decoded and reconstruction error is computed. Reconstruction error is then back-propagated through the network.

In practice, encoded distributions are *normal,* so the encoder can be trained to return the mean and covariance matrix that describe the Gaussian distributions. The reason an input is encoded as a distribution instead of a single point is because latent space regularization can be expressed very naturally as Gaussian. Distributions returned by the encoder are enforced to be close to a standard normal distribution by the VAE model.

For an excellent discussion of VAE modeling, peruse

https://towardsdatascience.com/understanding-variational-autoencoders-vaes-f70510919f73

Load Data

Load Fashion-MNIST train and test images as a single batch of NumPy arrays:

```
(x_train_fm, _), (x_test_fm, _) = tfds.as_numpy(
    tfds.load('fashion_mnist', split=['train','test'],
              batch_size=-1, as_supervised=True,
              try_gcs=True))
```

Inspect Data

Get train and test tensor shapes:

```
x_train_fm.shape, x_test_fm.shape
```

Get input shape for the encoder:

```
fmnist_shape = x_train_fm.shape[1:]
fmnist_shape
```

Scale

Scale feature images by dividing by the number of pixels in each image:

```
x_train_fds, x_test_fds = x_train_fm.astype(np.float32) / 255,\
                          x_test_fm.astype(np.float32) / 255
```

Create a Custom Layer to Sample Codings

The sampling layer for the encoder takes two inputs, namely, mean μ and log variance γ. It uses the tf.random.normal API to sample a random vector of the same shape as γ from the normal distribution with mean of zero (μ=0) and standard deviation of 1 (σ=1), multiplies the random vector by exp(γ/2), adds μ, and returns the result.

Create the sampling class as shown in Listing 9-6.

Listing 9-6. Sampling Class for the Encoder

```
class Sampling(tf.keras.layers.Layer):
  def call(self, inputs):
    mean, log_var = inputs
    return tf.random.normal(tf.shape(log_var)) *\
           tf.math.exp(log_var / 2) + mean
```

Create the VAE Model

Clear and seed:

```
tf.keras.backend.clear_session()
np.random.seed(0)
tf.random.set_seed(0)
```

Create the encoder as shown in Listing 9-7.

Listing 9-7. VAE Encoder

```
codings_size = 10

inputs = Input(shape=fmnist_shape)
z = Flatten()(inputs)
z = Dense(128, activation='relu')(z)
z = Dense(64, activation='relu')(z)
z = Dense(32, activation='relu')(z)
codings_mean = Dense(codings_size)(z)
codings_log_var = Dense(codings_size)(z)
codings = Sampling()([codings_mean, codings_log_var])
```

```
variational_encoder = Model(
    inputs=[inputs],
    outputs=[codings_mean, codings_log_var, codings])
```

We use the Functional API because the model isn't entirely sequential. The Dense layers output codings_mean μ and codings_log_var γ that both have the same inputs (i.e., the outputs of the second Dense layer). Both codings_mean and codings_log_var are then passed to the sampling layer. The variational_encoder has three outputs, namely, codings_mean, codings_log_var, and codings. But we only use the codings output.

Create the decoder as shown in Listing 9-8.

Listing 9-8. VAE Decoder

```
decoder_inputs = Input(shape=[codings_size])
x = Dense(32, activation='relu')(decoder_inputs)
x = Dense(64, activation='relu')(x)
x = Dense(128, activation='relu')(x)
x = Dense(28 * 28, activation='sigmoid')(x)
outputs = Reshape(fmnist_shape)(x)
variational_decoder = Model(
    inputs=[decoder_inputs], outputs=[outputs])
```

We could have used the Sequential API instead of the Functional API since it is really just a simple stack of layers. But we created the VAE decoder in this way to match the structure of the VAE encoder.

Build the VAE model:

```
_, _, codings = variational_encoder(inputs)
reconstructions = variational_decoder(codings)
variational_ae = Model(
    inputs=[inputs], outputs=[reconstructions])
```

Ignore the first two outputs because we only need the codings.

Compile and Train

Add Latent Loss and Reconstruction Loss to the Model

Compute latent loss as 1 plus codings_log_var minus the exponential of codings_log_var minus the square of codings_mean ($1 + \text{codings\_log\_var} - e^{\text{codings\_log\_var}} - \text{codings\_mean}^2$). Multiply this result by -0.5. Compute reconstruction loss as mean loss over all instances in the batch divided by 784 to ensure appropriate scale. We divide by 784 because images are 28 × 28 pixels.

As review, an autoencoder is a neural network that receives an image and reconstructs it by learning the input features from the image. The encoder creates a latent vector from the image. A *latent vector* is a compressed representation of the image. The latent vector is fed into a decoder, which reconstructs a facsimile of the input image from the latent vector.

A VAE also reconstructs a new image from the learned input features of an image. But the encoder creates a sample of latent vectors based on a Gaussian distribution. The decoder randomly draws latent vectors from the distribution created by the encoder to create new facsimile images. New images are better quality because they are based on random latent representations drawn from a Gaussian distribution.

We restrict the VAE with latent loss. *Latent loss* measures the loss of the latent vector against the Gaussian distribution created by the encoder. So latent loss penalizes the network based on how close or how far a latent vector is to or from the unit Gaussian distribution. The actual image loss is also measured to see how well the facsimile image matches the input image. When combining these two loss metrics, the network has to find the best trade-off between low latent loss (unit Gaussian distribution of latent vectors) and low image loss (high similarity between input and facsimile output images). The latent loss only evaluates to zero (perfect) when the mean is 0 and the standard deviation is 1, which is unit Gaussian.

Create the restricted VAE as shown in Listing 9-9.

Listing 9-9. Restricted VAE

```
latent_loss = -0.5 * tf.math.reduce_sum(
    1 + codings_log_var - tf.math.exp(codings_log_var) -\
    tf.math.square(codings_mean), axis=-1)

variational_ae.add_loss(
    tf.math.reduce_mean(latent_loss) / 784.)
```

Compile:

```
variational_ae.compile(
    loss='binary_crossentropy', optimizer='rmsprop',
    metrics=[rounded_accuracy])
```

Train:

```
vae_history = variational_ae.fit(
    x_train_fds, x_train_fds, epochs=10,
    batch_size=128,
    validation_data=(x_test_fds, x_test_fds))
```

Visualize training performance:

```
viz_history(vae_history)
```

Visualize Reconstructions

Inspect the shape of the test set:

```
x_test_fds.shape
```

Remove the *1* dimension for plotting purposes:

```
x_test_fds_imgs = tf.squeeze(x_test_fds)
x_test_fds_imgs.shape
```

Visualize:

```
show_reconstructions(
    variational_ae, x_test_fds_imgs, 5, reshape=True)
```

Generate New Images

Create a plotting function as shown in Listing 9-10.

Listing 9-10. Plotting Function for New Images

```
def plot_multiple_images(images, n_cols=None):
  n_cols = n_cols or len(images)
  n_rows = (len(images) - 1) // n_cols + 1
```

```
  if images.shape[-1] == 1:
    images = np.squeeze(images, axis=-1)
  plt.figure(figsize=(n_cols, n_rows))
  for index, image in enumerate(images):
    plt.subplot(n_rows, n_cols, index + 1)
    plt.imshow(image, cmap='binary')
    plt.axis('off')
```

Instead of generating reconstructions from image predictions, we create a random Gaussian set of codings and reconstruct images directly from the decoder.

Generate a few random codings, decode them, and plot the resulting images:

```
tf.random.set_seed(0)

codings = tf.random.normal(shape=[12, codings_size])
images = variational_decoder(codings).numpy()
plot_multiple_images(images, 4)
```

Create 12 random codings, decode them with the decoder, and plot them with the plotting function.

Perform semantic interpolation between the images as shown in Listing 9-11.

Listing 9-11. Semantic Interpolation Between the Codings

```
tf.random.set_seed(0)
np.random.seed(0)

codings_grid = tf.reshape(codings, [1, 3, 4, codings_size])
larger_grid = tf.image.resize(codings_grid, size=[5, 7])
interpolated_codings = tf.reshape(
    larger_grid, [-1, codings_size])
images = variational_decoder(interpolated_codings).numpy()
images.shape
```

With implicit models like a VAE, we interpolate between sampled points in latent space. We do so to match the distributional assumptions on the code vectors with the geometry of the interpolating paths. Otherwise, typical assumptions about the quality and semantics between coding points may not be justified.

Remove the *1* dimension for plotting purposes:

```
images = tf.squeeze(images)
```

Visualize as shown in Listing 9-12.

Listing 9-12. Visualize Interpolated Codings

```
plt.figure(figsize=(7, 5))
for index, image in enumerate(images):
  plt.subplot(5, 7, index + 1)
  if index%7%2==0 and index//7%2==0:
    plt.gca().get_xaxis().set_visible(False)
    plt.gca().get_yaxis().set_visible(False)
  else:
    plt.axis('off')
  plt.imshow(image, cmap='binary')
```

TFP Experiment

In this experiment, we fit a VAE using TensorFlow Probability Layers. *TensorFlow Probability (TFP)* is a library for probabilistic reasoning and statistical analysis in TensorFlow. As part of the TensorFlow ecosystem, TFP provides integration of probabilistic methods with deep networks, gradient-based inference using automatic differentiation, and scalability to large datasets and models with hardware acceleration (e.g., GPUs) and distributed computation.

TFP makes it easy to combine probabilistic models and deep learning on modern hardware (e.g., TPU, GPU). TFP is designed for data scientists, statisticians, ML researchers, and practitioners who want to encode domain knowledge to understand data and make predictions.

TFP includes a wide selection of probability distributions and bijectors. The *Bijector API* provides modular building blocks for constructing a broad class of probability distributions. Bijectors encapsulate the change of variables for a probability density. So they can be used to transform a distribution of tensors into another type of distribution. Simply, they turn one random outcome into another random outcome from a different distribution.

Since TFP inherits the benefits of TensorFlow, we can build, fit, and deploy a model using a single language throughout the lifecycle of model exploration and production. TFP is open source and available on GitHub.

Load Fashion-MNIST data:

```
fmnist, fmnist_info = tfds.load(
    name='fashion_mnist', try_gcs=True,
    with_info=True, as_supervised=False)
```

Create a function to preprocess the data:

```
def _preprocess(sample):
  image = tf.cast(sample['image'], tf.float32) / 255.
  image = image < tf.random.uniform(tf.shape(image))
  return image, image
```

Cast an image to float and scale. Continue by randomly casting image pixels to binary form as either *True* or *False* values. The tf.random.uniform API outputs random values from a uniform distribution. Notice that the function returns *image, image* rather than just image because Keras is intended for use with discriminative models with an *(example, label)* input format. Since the goal of a VAE is to recover the input x from x itself, the data pair is *(example, example)*.

Establish parameters for the input pipeline:

```
auto = tf.data.experimental.AUTOTUNE
BATCH_SIZE, SHUFFLE_SIZE = 256, int(10e3)
```

Create the train set:

```
train_tpl = (fmnist['train']
             .map(_preprocess)
             .batch(BATCH_SIZE)
             .prefetch(auto)
             .shuffle(SHUFFLE_SIZE))
```

Create the test set:

```
test_tpl = (fmnist['test']
            .map(_preprocess)
            .batch(BATCH_SIZE)
            .prefetch(auto))
```

Inspect a slice from an image from a batch:

```
for example in train_tpl.take(1):
  print (example[0][0][0])
  print (example[0].shape)
```

Notice that pixels are now either Boolean True or False!

Inspect the shape of a training batch:

```
for row in train_tpl.take(1):
  print (row[0].shape)
```

Create a TFP VAE Model

Create a TFP independent Gaussian distribution with no learned parameters. Latent variable z (encoded_size) is assigned 16 dimensions. Assign the distribution to prior as shown in Listing 9-13.

Listing 9-13. TFP Independent Gaussian Distribution

```
import tensorflow_probability as tfp

tfd = tfp.distributions
encoded_size = 16
prior = tfd.Independent(
    tfd.Normal(
        loc=tf.zeros(encoded_size), scale=1),
        reinterpreted_batch_ndims=1)
```

Notice that prior is independently normalized. A **prior** is an underlying assumption we have about the world. A common prior is that we assume a coin to be fair (50% heads and 50% tails). Common wisdom is that a prior is not always true, but most of the time it is true.

Get input shape and assign depth:

```
input_shape = fmnist_info.features['image'].shape
base_depth = 32
input_shape
```

Base depth is the base number of neurons.

Assign aliases for convenience:

```
tfpl = tfp.layers
tfd = tfp.distributions
leaky = tf.nn.leaky_relu
```

TFP VAE Encoder

Create an encoder with a full-covariance Gaussian distribution with mean and covariance matrices parameterized by the output of a neural network. TFP Layers enable construction of this complex encoder in an easy format.

Create the encoder as shown in Listing 9-14.

Listing 9-14. TFP VAE Encoder

```
tf.compat.v1.logging.set_verbosity(tf.compat.v1.logging.ERROR)

from tensorflow.keras.layers import InputLayer, Lambda

encoder = Sequential([
  InputLayer(input_shape=input_shape),
  Lambda(lambda x: tf.cast(x, tf.float32) - 0.5),
  Conv2D(base_depth, 5, strides=1, padding='same',
         activation=leaky),
  Conv2D(base_depth, 5, strides=2, padding='same',
         activation=leaky),
  Conv2D(base_depth * 2, 5, strides=1,
         padding='same', activation=leaky),
  Conv2D(base_depth * 2, 5, strides=2, padding='same',
         activation=leaky),
  Conv2D(4 * encoded_size, 7, strides=1, padding='valid',
         activation=leaky),
  Flatten(),
  Dense(tfpl.MultivariateNormalTriL.params_size(encoded_size),
        activation=None),
```

```
  tfpl.MultivariateNormalTriL(
      encoded_size,
      activity_regularizer=tfpl.KLDivergenceRegularizer(
          prior, weight=1.0))
])
```

The encoder is a normal Keras sequential model with convolutions and dense layers. But the output is passed to a TFP Layer (MultivariateNormalTril()), which transparently splits the activations from the final Dense layer into the parts needed to specify both the mean and the lower triangular covariance matrix. We use the tfpl helper MultivariateNormalTriL.params_size(encoded_size) to make the Dense layer output the correct number of activations. We also contribute a regularization term to the final loss to mitigate overfitting. Specifically, we add Kullback-Leibler (KL) divergence regularization between the encoder and the prior to the loss.

TFP VAE Decoder

Create the decoder as a pixel-independent Bernoulli distribution as shown in Listing 9-15.

Listing 9-15. TFP VAE Decoder

```
decoder = Sequential([
  InputLayer(input_shape=[encoded_size]),
  Reshape([1, 1, encoded_size]),
  Conv2DTranspose(2 * base_depth, 7, strides=1,
                  padding='valid', activation=leaky),
  Conv2DTranspose(2 * base_depth, 5, strides=1,
                  padding='same', activation=leaky),
  Conv2DTranspose(2 * base_depth, 5, strides=2,
                  padding='same', activation=leaky),
  Conv2DTranspose(base_depth, 5, strides=1,
                  padding='same', activation=leaky),
  Conv2DTranspose(base_depth, 5, strides=2,
                  padding='same', activation=leaky),
  Conv2DTranspose(base_depth, 5, strides=1,
                  padding='same', activation=leaky),
  Conv2D(filters=1, kernel_size=5, strides=1,
```

```
        padding='same', activation=None),
  Flatten(),
  tfpl.IndependentBernoulli(
     input_shape, tfd.Bernoulli.logits)
])
```

The form here is essentially the same as the encoder, but we use transposed convolutions to convert our latent representation of a 16-dimensional vector back into a 28 × 28 × 1 tensor. That final layer parameterizes the pixel-independent Bernoulli distribution.

Build the TFP VAE Model

Build the model:

```
from tensorflow.keras import Model

tpl_vae = Model(inputs=encoder.inputs,
                outputs=decoder(encoder.outputs[0]))
```

Check that inputs are as expected:

```
encoder.inputs
```

Compile and Train

Set a learning rate:

```
lr = 1e-3
lr
```

We set this learning rate based on trial and error experimentation. Feel free to try different learning rates.

Compile:

```
negloglik = lambda x, rv_x: -rv_x.log_prob(x)

tpl_vae.compile(
    optimizer=tf.optimizers.Adam(learning_rate=lr),
    loss=negloglik)
```

Our model is just a Keras model where outputs are defined as the composition of the encoder and the decoder. Since the encoder already added the Kullback-Leibler (KL) divergence term to the loss, we only need to specify the reconstruction loss (the first term of the ELBO). KL divergence is a measure of how one probability distribution is different from a second one. It is also called relative entropy.

ELBO (Evidence Lower BOund) is a lower bound on log p(x) (or the log probability of an observed data point). The first integral in the ELBO equation is the reconstruction term. It asks how likely we are to start at an image x, encode it to z, decode it, and get back the original x. The second term is the KL divergence term. It measures how close our encoder is to the value we assigned to the prior variable. We can think of this term as just trying to keep our encoder honest. If our encoder generates z samples that are too unlikely given our prior value, the objective is worse than if it generates z samples more typical of the prior value. Thus, the encoder should differ from the prior value only if the cost of doing so is outweighed by the benefit from the reconstruction term.

Train:

```
_ = tpl_vae.fit(train_tpl, epochs=15,
                validation_data=test_tpl)
```

Efficacy Test

Examine ten random digits:

```
x = next(iter(test_tpl))[0][:10]
xhat = tpl_vae(x)
assert isinstance(xhat, tfd.Distribution)
```

Create a function to display random images as shown in Listing 9-16.

Listing 9-16. Plotting Function for Random Images

```
def display_imgs(x, y=None):
  if not isinstance(x, (np.ndarray, np.generic)):
    x = np.array(x)
  plt.ioff()
  n = x.shape[0]
  fig, axs = plt.subplots(1, n, figsize=(n, 1))
  if y is not None:
```

```
    fig.suptitle(np.argmax(y, axis=1))
  for i in range(n):
    axs.flat[i].imshow(x[i].squeeze(),
                      interpolation='none',
                      cmap='gray')
    axs.flat[i].axis('off')
  plt.show()
  plt.close()
```

Visualize:

```
print('Originals:')
display_imgs(x)

print('Decoded Random Samples:')
display_imgs(xhat.sample())

print('Decoded Modes:')
display_imgs(xhat.mode())

print('Decoded Means:')
display_imgs(xhat.mean())
```

The Python sample() method returns a particular-length list of items chosen from a sequence. It is used for random sampling without replacement. The Python mode() method applies to nominal (nonnumeric) data. Remember that we converted pixel image data to Boolean True or False values. It is used to locate the central tendency of numeric or nominal data. The Python mean() method calculates the arithmetic average of a given list of numbers. To get clearer reconstruction images, run the model for many more epochs (to reduce loss).

Summary

In this chapter, we demonstrated three state-of-the-art autoencoder experiments. Each of these autoencoders produced more realistic renderings of input images than basic stacked autoencoders.

CHAPTER 10

Generative Adversarial Networks

Generative modeling is an unsupervised learning technique that involves automatically discovering and learning the regularities (or patterns) in input data so that a trained model can generate new examples that plausibly could have been drawn from the original dataset. A popular type of generative model is a generative adversarial network. **Generative adversarial networks** (GANs) are generative models that create new data instances that resemble the training data.

GANs frame the problem as a supervised learning problem with two submodels, namely, a generator model and discriminator model. The *generator* model is trained to generate new examples. The *discriminator* model attempts to classify examples as either real (from the domain) or fake (generated). The domain represents images from the original training set. The two models are trained together in a zero-sum adversarial game until the discriminator model is fooled about half the time. The result of training is generation of plausible examples.

GANs excel at realistic examples in image-to-image translation tasks such as translating photos of summer to winter or day to night. They have also been successful at generating photorealistic photos of objects, scenes, and people that even humans cannot tell are fake. GANs can create images that look like photographs of human faces even though the faces don't belong to any real person.

We demonstrate a GAN with a code example. We also demonstrate Deep Convolutional GANs (DCGANs) with code examples. A GAN uses a feedforward network to learn, while a Deep Convolutional GAN uses a convolutional network to learn.

Notebooks for chapters are located at the following URL:

`https://github.com/paperd/deep-learning-models`

D. Paper, *State-of-the-Art Deep Learning Models in TensorFlow*, https://doi.org/10.1007/978-1-4842-7341-8_10

We begin with a GAN experiment. We continue with two Deep Convolutional GAN (DCGAN) experiments. A GAN generates facsimile images from input images. A DCGAN does the same, but generates much more realistic facsimiles.

Begin setting up the Colab ecosystem by importing the main TensorFlow library and instantiating the GPU.

Import the TensorFlow Library

Import the library and alias it as **tf**:

```
import tensorflow as tf
```

GPU Hardware Accelerator

As a convenience, we include the steps to enable the GPU in a Colab notebook:

1. Click *Runtime* in the top-left menu
2. Click *Change runtime type* from the drop-down menu.
3. Choose *GPU* from the *Hardware accelerator* drop-down menu.
4. Click *Save*.

Verify that the GPU is active:

```
tf.__version__, tf.test.gpu_device_name()
```

If '/device:GPU:0' is displayed, the GPU is active. If '..' is displayed, the regular CPU is active.

Note If you get the error **NAME 'TF' IS NOT DEFINED**, re-execute the code to import the TensorFlow library!

GAN Experiment

A GAN can generate a high level of realism in the new data instances that it creates by pairing a generator that learns to produce the target output with a discriminator that learns to distinguish true data from the output of the generator. The generator tries to fool the discriminator, and the discriminator tries to keep from being fooled.

During training, the generator and discriminator have opposite goals. The discriminator tries to distinguish fake images from real ones. The generator tries to produce images that look real enough to trick the discriminator.

We begin with a simple feedforward GAN to show you how one can be trained on a simple dataset.

Load Data

Load Fashion-MNIST as NumPy arrays:

```
import tensorflow_datasets as tfds

x_train_img, _ = tfds.as_numpy(
    tfds.load('fashion_mnist', split='train',
              batch_size=-1, as_supervised=True,
              try_gcs=True, shuffle_files=True))
```

Since our experiment is *unsupervised*, we only need training images.

Get number of examples:

```
len(x_train_img)
```

Scale

Scale examples:

```
import numpy as np

images = x_train_img.astype(np.float32) / 255
images.shape
```

Verify scaling:

```
x_train_img[0][0], images[0][0]
```

Scaling works as expected.

Build the GAN

Get input shape:

```
in_shape = images.shape[1:]
in_shape
```

Clear and seed:

```
tf.keras.backend.clear_session()
np.random.seed(0)
tf.random.set_seed(0)
```

Import libraries:

```
from tensorflow.keras.models import Sequential
from tensorflow.keras.layers import Dense, Flatten,\
  Reshape
```

Build the generator as shown in Listing 10-1.

Listing 10-1. The Generator

```
codings_size = 30

generator = Sequential([
  Dense(32, activation='selu', input_shape=[codings_size]),
  Dense(64, activation='selu'),
  Dense(128, activation='selu'),
  Dense(28 * 28, activation='sigmoid'),
  Reshape(in_shape)
])
```

As we know, a *generator* is used to generate new plausible examples from the problem domain. It takes a random distribution as input (typically Gaussian) and outputs some data (typically an image). The random inputs can be thought of as the latent representations (or codings) of the image to be generated. So generators have the same functionality as decoders in a variational autoencoder and can be used in the same way to generate new images. However, they are trained very differently.

The generator learns to create fake data by incorporating feedback from the discriminator. It learns to make the discriminator classify its output as real.

Generator training requires tighter integration between the generator and the discriminator than discriminator training requires. The portion of the GAN that trains the generator includes random input, the generator network, discriminator output, and generator loss. The generator network transforms random input into a data instance. The discriminator network classifies the generated data. Generator loss penalizes the generator for failing to fool the discriminator.

Now, build the discriminator as shown in Listing 10-2.

Listing 10-2. The Discriminator

```
discriminator = Sequential([
  Flatten(input_shape=in_shape),
  Dense(128, activation='selu'),
  Dense(64, activation='selu'),
  Dense(32, activation='selu'),
  Dense(1, activation='sigmoid')
])
```

As we also know, a *discriminator* is used to classify examples as real (from the domain) or fake (generated from the generator). It takes either a fake image from the generator or a real image from the training set as input and guesses whether the input image is fake or real. The discriminator is a regular binary classifier (real or fake images). It takes an image as input and ends with a Dense layer containing a single unit.

Create the model:

```
gan = Sequential([generator, discriminator])
```

Compile the Discriminator Model

Since the discriminator is naturally a binary classifier (fake or real images), we naturally use binary cross-entropy loss. Since the generator is only trained through the GAN model, we *don't need* to compile it. The GAN model is also a binary classifier, so it can use the same loss function. Importantly, the discriminator should not be trained during the second phase. So we make it non-trainable before compiling the GAN model.

```
discriminator.compile(
    loss='binary_crossentropy', optimizer='rmsprop')
discriminator.trainable = False
gan.compile(loss='binary_crossentropy', optimizer='rmsprop')
```

Training data fed to the discriminator comes from two sources - real and fake data. For our experiment, real data instances are Fashion-MNIST images. The discriminator uses these instances as positive examples during training. Fake data instances are created by the generator. The discriminator uses these instances as negative examples during training.

During discriminator training, the generator does not train. Its weights remain constant while it produces examples for the discriminator to train on.

The discriminator connects to two loss functions - generator and discriminator. During discriminator training, the discriminator ignores the generator loss and just uses the discriminator loss. The discriminator classifies both real data and fake data from the generator. Discriminator loss penalizes the discriminator for misclassifying a real instance as fake or a fake instance as real. The discriminator updates its weights through back-propagation from the discriminator loss through the discriminator network.

Build the Input Pipeline

Build the pipeline with batch size of 32:

```
batch_size = 32

dataset = tf.data.Dataset.from_tensor_slices(
    images).shuffle(1000)
dataset = dataset.batch(
    batch_size, drop_remainder=True).prefetch(1)
```

Create a Custom Loop for Training

Since the training loop is unusual, we can't use the regular fit method. Instead we create a custom loop that accepts a dataset to iterate through images. Training is unusual because it is not sequential. The discriminator is trained on real and fake images fed by the generator in phase 1. The generator is trained on what the discriminator produces from what it learned in phase 2. So there are feedback loops in the training process.

Create a training loop function as shown in Listing 10-3.

Listing 10-3. Custom Training Loop Function

```
def train_gan(gan, dataset, batch_size,
              codings_size, n_epochs=50):
  generator, discriminator = gan.layers
  for epoch in range(n_epochs):
    print('Epoch {}/{}'.format(epoch + 1, n_epochs))
    for X_batch in dataset:
      # phase 1 - training the discriminator
      noise = tf.random.normal(
          shape=[batch_size, codings_size])
      generated_images = generator(noise)
      X_fake_and_real = tf.concat(
          [generated_images, X_batch], axis=0)
      y1 = tf.constant([[0.]] * batch_size + [[1.]] * batch_size)
      discriminator.trainable = True
      discriminator.train_on_batch(X_fake_and_real, y1)
      # phase 2 - training the generator
      noise = tf.random.normal(
          shape=[batch_size, codings_size])
      y2 = tf.constant([[1.]] * batch_size)
      discriminator.trainable = False
      gan.train_on_batch(noise, y2)
    plot_multiple_images(generated_images, 8)
    plt.show()
```

Each training iteration is divided into two phases. The first phase trains the discriminator. A batch of real images is sampled from the training set, and an equal number of fake images is produced by the generator. Labels are set to 0 for fake images and 1 for real images. The discriminator is trained on the labeled batch for one step using binary cross-entropy loss. Importantly, back-propagation only optimizes the weights of the discriminator during this phase. The second phase trains the generator. The generator produces another batch of fake images, and the discriminator distinguishes between fake and real images. No real images are added to the batch, and all labels are set to 1 to indicate they are real (though they are not real). The idea is that the generator produces images that the discriminator incorrectly believes are real! Crucially, weights of the discriminator are frozen during this step so back-propagation only affects the weights of the generator.

Technically speaking, phase 1 feeds Gaussian noise to the generator to produce fake images. Target *y1* is set to 0 for fake and 1 for real images. The discriminator is then trained on the batch. Phase 2 feeds the GAN some Gaussian noise. The generator starts by producing fake images. The discriminator then tries to guess if images are fake or real. The idea is to make the discriminator believe that the fake images are real, so target *y2* is set to 1.

Humans can easily ignore noise, but ML algorithms struggle. Small, human-imperceptible pixel changes can dramatically alter a neural network's ability to make an accurate prediction. Research has shown that *noise* and *Gaussian blurring* show near-immediate smoothing on tested models. **Gaussian blurring** (also known as Gaussian smoothing) is the result of blurring an image by a Gaussian function. It is widely used in ML to reduce image noise and reduce detail.

For an excellent resource on adding noise, peruse

https://blog.roboflow.com/why-to-add-noise-to-images-for-machine-learning/

Create a function for plotting as shown in Listing 10-4.

Listing 10-4. Plotting Function for Generated Images

```
import matplotlib.pyplot as plt

def plot_multiple_images(images, n_cols=None):
  n_cols = n_cols or len(images)
  n_rows = (len(images) - 1) // n_cols + 1
  if images.shape[-1] == 1:
```

```
    images = np.squeeze(images, axis=-1)
  plt.figure(figsize=(n_cols, n_rows))
  for index, image in enumerate(images):
    plt.subplot(n_rows, n_cols, index + 1)
    plt.imshow(image, cmap='binary')
    plt.axis('off')
```

Generate images with the untrained generator:

```
tf.random.set_seed(0)
np.random.seed(0)

noise = tf.random.normal(shape=[batch_size, codings_size])
generated_images = generator(noise)
plot_multiple_images(generated_images, 8)
```

As expected, generated images are not too impressive. Of course, we have yet to train the model!

Train the GAN

Train the GAN for a few epochs:

```
n = 5

train_gan(gan, dataset, batch_size, codings_size, n_epochs=n)
```

Generated images are better than no training at all, but still they are not too impressive! We trained for 50 epochs with little improvement. That is, image quality stops improving. We only train for a few epochs here because it takes quite a bit of time. Feel free to train on even more epochs, but you might want to take a lunch break because training will take a lot of time.

Note If images look like blobs, restart the Colab notebook runtime and rerun the experiment.

Generate images from the trained GAN:

```
np.random.seed(0)
tf.random.set_seed(0)

noise = tf.random.normal(shape=[batch_size, codings_size])
generated_images = generator(noise)
plot_multiple_images(generated_images, 8)
```

Add some Gaussian noise and use the generator model to generator some images. There are 32 images generated because batch size is 32. We add Gaussian noise to compensate for training the GAN with Gaussian noise.

DCGAN Experiment with Small Images

Using CNN architectures with GANs can produce much better results than simple feedforward networks with GANs. A **Deep Convolutional generative adversarial network** (DCGAN) is a GAN with convolutional neural networks as its generator and discriminator. Instead of using a feedforward network for the generator and discriminator, we substitute a CNN. The DCGAN architecture achieves superior performance with image processing tasks in many cases.

Create the Generator

Import libraries:

```
from tensorflow.keras.layers import BatchNormalization,\
  Conv2D, Conv2DTranspose, LeakyReLU, Dropout
```

Create the generator as shown in Listing 10-5.

Listing 10-5. DCGAN Generator

```
codings_size = 100

dc_generator = Sequential([
  Dense(7 * 7 * 128, input_shape=[codings_size]),
  Reshape([7, 7, 128]),
  BatchNormalization(),
```

```
    Conv2DTranspose(
        64, kernel_size=5, strides=2, padding='SAME',
        activation='selu'),
    BatchNormalization(),
    Conv2DTranspose(
        1, kernel_size=5, strides=2, padding='SAME',
        activation='tanh'),
])
```

Begin by feeding the generator small tensors (7 × 7 pixels) projected to 128 dimensions resulting in a 6,272-dimensional space. Multiply 7 × 7 × 128 to get the dimensional space. The goal is to increase tensor image size to match Fashion-MNIST images at 28 × 28 and decrease depth to 1 to match channel size. We created this network through trial and error experimentation. Experiment with the initial tensor size, but be aware that doing so changes the neuron calculations for both the generator and discriminator.

Specifically, the generator accepts codings of size 100 and projects them to 6,272 dimensions (7 × 7 × 128). The generator then reshapes the projection to a 7 × 7 × 128 tensor, which is batch-normalized and fed to a transposed convolutional layer with a stride of 2. The stride up-samples the tensor to 14 × 14 because it doubles the 7 × 7 dimension. The layer also reduces the tensor's depth from 128 to 64. The result is batch-normalized again and fed to another transposed convolutional layer with a stride of 2. The stride up-samples it to 28 × 28 because it doubles the 14 × 14 dimension. The layer also reduces the tensor's depth from 64 to 1. The output tensor has shape (28, 28, 1), which is the goal because Fashion-MNIST images have shape 28 × 28 × 1. Since the final layer uses *tanh* activation, outputs are reshaped between the range -1 and 1.

The hyperbolic tangent activation function is also referred to as the tanh function. It is very similar to the sigmoid activation function and even has the same S shape. But the function takes any real value as input and outputs values in the range -1 to 1. We used tanh instead of sigmoid for the output layer because it performed better in this case.

Generate images with the untrained generator:

```
tf.random.set_seed(0)
np.random.seed(0)

noise = tf.random.normal(shape=[batch_size, codings_size])
generated_images = dc_generator(noise)
plot_multiple_images(generated_images, 8)
```

Create the Discriminator

Create the discriminator as shown in Listing 10-6.

Listing 10-6. DCGAN Discriminator

```
dc_discriminator = Sequential([
  Conv2D(64, kernel_size=5, strides=2, padding='SAME',
         activation=LeakyReLU(0.2),
         input_shape=[28, 28, 1]),
  Dropout(0.4),
  Conv2D(128, kernel_size=5, strides=2, padding='SAME',
         activation=LeakyReLU(0.2)),
  Dropout(0.4),
  Flatten(),
  Dense(1, activation='sigmoid')
])
```

The discriminator looks like a regular CNN for binary classification (the final Dense layer is 1), except it uses strides to down-sample instead of max pooling layers.

Create the DCGAN

Create the DCGAN from the DCGAN generator and discriminator:

```
dcgan = Sequential([dc_generator, dc_discriminator])
```

Compile the Discriminator Model

Since the discriminator is naturally a binary classifier (fake or real images), we naturally use binary cross-entropy loss. Since the generator is only trained through the DCGAN model, we *don't need* to compile it. The DCGAN model is also a binary classifier, so it can use the same loss function. Importantly, the discriminator should not be trained during the second phase. So we make it non-trainable before compiling the DCGAN model.

Compile the DCGAN:

```
dc_discriminator.compile(
    loss='binary_crossentropy', optimizer='rmsprop')
dc_discriminator.trainable = False
dcgan.compile(loss='binary_crossentropy', optimizer='rmsprop')
```

Reshape

Since outputs range from -1 to 1 due to tanh activation in the final layer of the generator, rescale the training set to the same range:

```
images_dcgan = tf.reshape(
    images, [-1, 28, 28, 1]) * 2. - 1.
```

Build the Input Pipeline

Build the pipeline with batch size of 32:

```
batch_size = 32
dataset = tf.data.Dataset.from_tensor_slices(images_dcgan)
dataset = dataset.shuffle(1000)
dataset = dataset.batch(
    batch_size, drop_remainder=True).prefetch(1)
```

Train

Clear and seed:

```
tf.keras.backend.clear_session()
np.random.seed(0)
tf.random.set_seed(0)
```

Train for a few epochs:

```
n = 5

train_gan(
    dcgan, dataset, batch_size, codings_size, n_epochs=n)
```

Wow! Much better!

Generate Images with the Trained Generator

Use the trained DC generator to make images:

```
tf.random.set_seed(0)
np.random.seed(0)

noise = tf.random.normal(shape=[batch_size, codings_size])
generated_images = dc_generator(noise)
plot_multiple_images(generated_images, 8)
```

DCGAN Experiment with Large Images

A very simple Deep Convolutional GAN worked well with Fashion-MNIST. But images in this dataset are small and grayscale. Let's see how well a Deep Convolutional GAN works with the *rock_paper_scissors* dataset, which contains large color images of hands playing the rock-paper-scissors game.

Inspect Metadata

Load the dataset to inspect its metadata:

```
rps, info = tfds.load('rock_paper_scissors', with_info=True,
                      split='train', try_gcs=True)
```

Inspect the metadata object:

```
info
```

Get class labels and number of classes:

```
num_classes = info.features['label'].num_classes
classes = info.features['label'].names
classes, num_classes
```

Visualize some examples:

```
fig = tfds.show_examples(rps, info)
```

Load Data for Training

Load train and test images as NumPy arrays:

```
(x_train_img, _), (x_test_img, _) = tfds.as_numpy(
    tfds.load('rock_paper_scissors', split=['train','test'],
              batch_size=-1, as_supervised=True,
              try_gcs=True))
```

Inspect shapes:

```
for element in range(10):
  print (x_train_img.shape)
```

The dataset contains 2,520 300 × 300 × 3 images. Since images are the same size, we don't have to resize them. However, we resize images for another reason explained later in the chapter.

Massage the Data

Shrink the dataset to a power of 2 to guarantee equal batches:

```
x_train_rps = x_train_img[:2048]
len(x_train_rps)
```

When we tried to train with the original dataset size, we got an error. So we created equal batches to eradicate the error.

Create a function to reformat images:

```
IMAGE_RES = 256

def format_image(image):
  image = tf.image.resize(image, (IMAGE_RES, IMAGE_RES))/255.0
  return image
```

Although we don't have to resize images, we do so to make it easier to create a model with an appropriate number of neurons at each layer of the network. We mentioned earlier in the chapter that all images are the same 300 × 300 pixel size. But we resize to a power of 2 for convenience.

A generator begins by projecting relatively small images onto many dimensions with a large depth size and gradually increases image size and decreases depth size until it outputs the image size we want. So it's easier to work with values divisible by and multiplicative of 2. Deep learning models also perform better when the number of neurons in a model is based on the power of 2.

Build the Input Pipeline

Create tensors from feature images:

```
train_slice = tf.data.Dataset.from_tensor_slices(x_train_rps)
```

Transform images for peak performance:

```
BATCH_SIZE = 32
SHUFFLE_SIZE = 500

train_rps = (train_slice.
             shuffle(SHUFFLE_SIZE).
             map(format_image).
             batch(BATCH_SIZE).
             cache().
             prefetch(1))
train_rps
```

Visualize some examples from a batch as shown in Listing 10-7.

Listing 10-7. Visualize Examples from a Batch

```
plt.figure(figsize=(10, 10))
for images in train_rps.take(1):
  for i in range(9):
    ax = plt.subplot(3, 3, i + 1)
    plt.imshow(images[i])
    plt.axis('off')
```

We don't show labels because we don't need them for our unsupervised experiment.

Build the Model

Clear and seed:

```
tf.keras.backend.clear_session()
np.random.seed(0)
tf.random.set_seed(0)
```

Create the generator as shown in Listing 10-8.

Listing 10-8. The Generator

```
codings_size = 100

gencolor = Sequential([
  Dense(32 * 32 * 256, input_shape=[codings_size]),
  Reshape([32, 32, 256]),
  BatchNormalization(),
  Conv2DTranspose(128, kernel_size=5, strides=2, padding='SAME',
                  activation='selu'),
  BatchNormalization(),
  Conv2DTranspose(64, kernel_size=5, strides=2, padding='SAME',
                  activation='selu'),
  BatchNormalization(),
  Conv2DTranspose(3, kernel_size=5, strides=2, padding='SAME',
                  activation='tanh'),
])
```

The number of neurons is based on the power of 2 with one exception. The final layer contains three neurons to represent the number of classes.

Since we want to generate large color images, begin by feeding the generator a bigger (32 × 32) tensor projected to 256 dimensions. So the resulting dimensional space is a 262,144. Multiply 32 by 32 by 256 to get this value.

The generator reshapes the projection to a 32 × 32 × 256 tensor, which is batch-normalized and fed to a transposed convolutional layer with a stride of 2. The stride up-samples the tensor to 64 × 64 because it doubles the 32 × 32 dimension. The layer also reduces the tensor's depth from 256 to 128. The result is batch-normalized again and fed to another transposed convolutional layer with a stride of 2. The stride up-samples it to 128 × 128 because it doubles the 64 × 64 dimension. The layer also reduces the tensor's depth from 128 to 64. The result is batch-normalized again and fed to another transposed convolutional layer with a stride of 2. The stride up-samples it to 256 × 256 because it doubles the 128 × 128 dimension. The layer also reduces the tensor's depth from 64 to 3.

The output tensor has shape (256, 256, 3), which is our goal because we want them to be the original 256 × 256 × 3 shape. Since the final layer uses tanh activation, outputs are reshaped between the range –1 and 1.

Note Hopefully you can see why we use powers of 2 to layer the model. We start our generator with a large dimensional space and, layer by layer, we reduce tensor dimensions back to the original shape of images from our dataset. Up-sampling and reducing tensors by 2 is very easy mathematically.

Inspect:

```
gencolor.summary()
```

Create the discriminator as shown in Listing 10-9.

Listing 10-9. The Discriminator

```
discolor = Sequential([
  Conv2D(64, kernel_size=5, strides=2, padding='SAME',
         activation=LeakyReLU(0.2),
         input_shape=[256, 256, 3]),
  Dropout(0.4),
  Conv2D(128, kernel_size=5, strides=2, padding='SAME',
         activation=LeakyReLU(0.2)),
  Dropout(0.4),
  Flatten(),
  Dense(1, activation='sigmoid')
])
```

The discriminator accepts 256 × 256 × 3 images, down-samples tensors to 128 × 128, and increases depth from 32 to 64. Tensors are fed to a transposed convolutional layer, which down-samples tensors to 64 × 64 and increases depth to 128. The result is flattened to a 524,288-dimensional space (64 × 64 × 128). Since we are conducting binary classification, the final Dense layer includes only *one* neuron.

The generator up-samples tensors to activate neurons in a higher-dimensional space. The discriminator down-samples tensors back to the original image size from the data for effective feature classification.

Inspect:

```
discolor.summary()
```

Create the DCGAN:

```
dcgan_color = Sequential([gencolor, discolor])
```

Compile the Discriminator Model

Since the discriminator is naturally a binary classifier (fake or real images), we naturally use binary cross-entropy loss. Since the generator is only trained through the DCGAN model, we *don't need* to compile it. The DCGAN model is also a binary classifier, so it can use the same loss function. Importantly, the discriminator should not be trained during the second phase. So we make it non-trainable before compiling the DCGAN model:

```
discolor.compile(
    loss='binary_crossentropy', optimizer='rmsprop')
discolor.trainable = False
dcgan_color.compile(
    loss='binary_crossentropy', optimizer='rmsprop')
```

Rescale

Create a function to rescale:

```
def rescale(image):
  image = tf.math.multiply(image, image * 2. -1)
  return image
```

Since outputs range from -1 to 1 due to tanh activation in the final layer of the generator, rescale the training set to the same range.

Rescale images:

```
train_color = train_rps.map(rescale)
```

Train Model and Generate Images

Create a function to plot images as shown in Listing 10-10.

Listing 10-10. Plotting Function for Generated Images

```
def plot_color(images, n_cols=None):
  n_cols = n_cols or len(images)
  n_rows = (len(images) - 1) // n_cols + 1
  images = np.clip(images, 0, 1)
  if images.shape[-1] == 1:
    images = np.squeeze(images, axis=-1)
  plt.figure(figsize=(n_cols, n_rows))
  for index, image in enumerate(images):
    plt.subplot(n_rows, n_cols, index + 1)
    plt.imshow(image, cmap='binary')
    plt.axis('off')
```

Create a training loop function to train the DCGAN as shown in Listing 10-11.

Listing 10-11. Training Loop Function

```
def train_gan(
    gan, dataset, batch_size, codings_size, n_epochs=50):
  generator, discriminator = gan.layers
  for epoch in range(n_epochs):
    print('Epoch {}/{}'.format(epoch + 1, n_epochs))
    for X_batch in dataset:
      # phase 1 - training the discriminator
      noise = tf.random.normal(
          shape=[batch_size, codings_size])
      generated_images = generator(noise)
      X_fake_and_real = tf.concat(
          [generated_images, X_batch], axis=0)
      y1 = tf.constant(
          [[0.]] * batch_size + [[1.]] * batch_size)
      discriminator.trainable = True
      discriminator.train_on_batch(X_fake_and_real, y1)
```

```
        # phase 2 - training the generator
        noise = tf.random.normal(
            shape=[batch_size, codings_size])
        y2 = tf.constant([[1.]] * batch_size)
        discriminator.trainable = False
        gan.train_on_batch(noise, y2)
    plot_color(generated_images, 8)
    plt.show()
```

Train the DCGAN:

```
train_gan(
    dcgan_color, train_color, BATCH_SIZE, codings_size, 10)
```

With just a few epochs, we can see that the generated images begin to represent rock, paper, and scissors hands. But the generator and discriminator models we built are very simple. To generate realistic images, we need to create a much more robust model.

Summary

In this chapter, we built a GAN with a feedforward network with mediocre generated facsimiles. With a DCGAN, generated facsimiles were much more realistic for Fashion-MNIST. For rock_paper_scissors, facsimiles weren't great. We could try training the model for more epochs, but this takes a lot of time and may take more memory than your computer can handle.

CHAPTER 11

Progressive Growing Generative Adversarial Networks

GANs are effective at generating crisp synthetic images, but are limited in size to about 64 × 64 pixels. A **Progressive Growing GAN** is an extension of the GAN that enables training generator models to generate large high-quality images up to about 1024 × 1024 pixels (as of this writing). The approach has proven effective at generating high-quality synthetic faces that are startlingly realistic.

The key innovation of Progressive Growing GANs is the incremental increase in the size of images output by the generator. By generating small images at the beginning of training and gradually adding convolutional layers to both generator and discriminator models, larger and larger images with finer and finer resolution are generated.

The technique begins with a low-resolution vector as input. It continues by progressively growing the generator and discriminator by adding new layers so the model can increasingly learn fine details during the training process. The technique also speeds and stabilizes training while generating images of unprecedented quality.

Latent Space Learning

Latent space is a representation of compressed data in which similar data points are closer together in space. Latent space is useful for learning data features and finding simpler representations of data for analysis. The idea behind creating a latent space is to compress reality so vector math works. Latent space can also be referred to as *latent vector space*.

D. Paper, *State-of-the-Art Deep Learning Models in TensorFlow*, https://doi.org/10.1007/978-1-4842-7341-8_11

When we observe our world, we see a vast landscape of pixels (or *observed pixel space*). But how can we hope to learn from such a giant canvas of data? One solution is to create a latent space that compresses an observed pixel space into manageable pixel images. For instance, to teach a model to learn human faces, we begin by taking (or using existing) pictures of them. We then convert the pictures into a set of pixel images. So each face is represented by a set of pixels. Now that we have a latent representation, we can apply calculus and vector arithmetic on image pixels to teach a learning model the essence of human faces (at least from a pixel image perspective). Substantively, the latent space for our experiment is a compressed pixel space representation culled from observed pixel spaces of actual human faces.

In an observed pixel space, there may be no immediate similarity between any two images. But mapping a pixel space to a latent space compresses images to be much closer together so we can more easily learn about such images.

A generative model in the GAN architecture learns to map points in a latent space for image generation. It takes a point from the latent space as input and applies vector arithmetic to generate a new image. A series of points can also be created on a linear path between two points in the latent space to create multiple generated images. In practice, a generative model effectively uses its latent space representation to interpolate between points in its latent space with the goal of deriving meaningful and targeted effects from its generated images. But the latent space only has meaning as it applies to the generative model being trained. That is, every learning experiment has its own latent space.

Notebooks for chapters are located at the following URL:

`https://github.com/paperd/deep-learning-models`

We present two Progressive Growing GAN experiments. The first experiment generates images from a pre-trained model. The second experiment creates a custom training loop to learn a target image from an initial generated image. Begin setting up the Colab ecosystem by importing the main TensorFlow library and instantiating the GPU.

Import the TensorFlow Library

Import the library and alias it as **tf**:

```
import tensorflow as tf
```

Aliasing the TensorFlow library as tf is common practice.

GPU Hardware Accelerator

As a convenience, we include the steps to enable the GPU in a Colab notebook:

1. Click *Runtime* in the top-left menu.
2. Click *Change runtime type* from the drop-down menu.
3. Choose *GPU* from the *Hardware accelerator* drop-down menu.
4. Click *Save*.

Verify that the GPU is active:

```
tf.__version__, tf.test.gpu_device_name()
```

If '/device:GPU:0' is displayed, the GPU is active. If '..' is displayed, the regular CPU is active.

Note If you get the error **NAME 'TF' IS NOT DEFINED**, re-execute the code to import the TensorFlow library!

Create Environment for Experiments

Both experiments expect packages, libraries, and functions for image and animation display.

Install Packages for Creating Animations

Install imageio, scikit-image, and TensorFlow docs packages to enable animation creation for the experiments:

```
!pip -q install imageio
!pip -q install scikit-image
!pip install -q git+https://github.com/tensorflow/docs
```

Install Libraries

Install the logging module on top of standard logging:

```
from absl import logging
```

The logging module is from the Abseil Python package, which provides various libraries for building Python applications. The module defines functions and classes that allow designers to build a flexible event logging system for applications and libraries. The key benefit of the logging API is that all Python modules can participate in logging. So our application log can include its own messages integrated with messages from third-party modules.

Install image processing libraries:

```
import imageio
import PIL.Image
import matplotlib.pyplot as plt
from IPython import display
from skimage import transform
```

The *imageio* library provides an easy interface to read and write a wide range of image data including animated images, volumetric data, and scientific formats. The *PIL.Image* module is used to represent a Python Imaging Library (PIL) image. The library adds support for opening, manipulating, and saving many different image file formats. The *plt* library is used for displaying images. The *display* module is a public API for display tools in IPython. *IPython* is an interactive shell built with Python. The *transform* module is used for image processing.

Import other requisite libraries:

```
import numpy as np
import tensorflow_hub as hub
from tensorflow_docs.vis import embed
import time
```

NumPy is a Python library used for working with scalars, arrays, and matrices. The *hub* module allows access to TensorFlow Hub, which is a repository of trained machine learning models. The *embed* module is used to embed animation in a notebook. The *time* library provides many ways of representing time in code such as objects, numbers, and strings.

Create Functions for Image Display

Create a function to display a PIL image:

```
def display_image(image):
  image = tf.constant(image)
  image = tf.image.convert_image_dtype(image, tf.uint8)
  return PIL.Image.fromarray(image.numpy())
```

Create a function to display an image:

```
def show_image(image):
  plt.imshow(image)
  plt.axis('off')
  plt.show()
```

The *imshow()* function is used to display data as an image.

Create a function to display animation as shown in Listing 11-1.

Listing 11-1. Function to Display Animation

```
def animate(images):
  images = np.array(images)
  converted_images = np.clip(
      images * 255, 0, 255).astype(np.uint8)
  imageio.mimsave('./animation.gif', converted_images)
  return embed.embed_file('./animation.gif')
```

Create Latent Space Dimensions

Latent space is useful for learning data features and finding simpler representations of data for analysis. Humans have an understanding of a broad range of topics and the events belonging to those topics. Latent space aims to provide a similar understanding for a computer model through a quantitative spatial representation. So latent space is really just a measurable spatial representation of a compressed reality for a specific training experiment.

Compressing a dataset into latent space helps a model better understand the observed data because the model deals with much smaller variations than it would with the entire dataset. So the model learns from a smaller space than if it had to learn from the actual observed pixel space.

The terms high dimensional and low dimensional help define how specific or general the kinds of features we want our latent space to learn and represent. *High-dimensional* latent space is sensitive to more specific features of the input data, but can sometimes lead to overfitting when there isn't sufficient training data. *Low-dimensional* latent space aims to capture the most important features (or aspects) required to learn and represent the input data.

Set a high-dimensional latent space of 512 because the pre-trained model we use for this experiment was trained on this latent space:

```
latent_dim = 512
```

The pre-trained model maps from a 512-dimensional latent space to images. We can retrieve latent dimensions from the *module.structured_input_signature* method if we don't know beforehand the module we are using. We show how to do this later in the chapter.

Set Verbosity for Error Logging

To see logging errors:

```
logging.set_verbosity(logging.ERROR)
```

Image Generation Experiment

We use the progan-128 pre-trained model to generate realistic images of human faces. The *progan-128* model is a Progressive GAN trained on CelebA (CelebFaces Attributes Dataset) for 128 × 128 pixel images. It maps from a 512-dimensional latent space to images. During training, the latent space vectors are sampled from a normal distribution.

The module takes a tensor (Tensor(tf.float32, shape=[?, 512]) that represents a batch of latent vectors as input and outputs a tensor (Tensor(tf.float32, shape=[?, 128, 128, 3]) that represents a batch of RGB images. The original model is trained on a GPU for 636,801 steps with a batch size 16.

CelebA is a large-scale face attributes dataset containing more than 200,000 celebrity images. Each image has 40 attribute annotations. Images in this dataset cover large pose variations and background clutter. CelebA also has large diversities, large quantities, and rich annotations including

* 10,177 unique celebrity images

* 202,599 face images

* 5 landmark locations

* 40 binary attribute annotations per image

CelebA can be employed as the training and test sets for computer vision tasks such as face attribute recognition, face detection, landmark (or facial part) localization, and face editing and synthesis.

Create a Function to Interpolate a Hypersphere

Create a function to interpolate the space between vectors in the latent space as shown in Listing 11-2.

Listing 11-2. Function to Interpolate the Hypersphere

```
def interpolate_hypersphere(v1, v2, num_steps):
  v1_norm = tf.norm(v1)
  v2_norm = tf.norm(v2)
  v2_normalized = v2 * (v1_norm / v2_norm)
  vectors = []
  for step in range(num_steps):
    interpolated =\
      v1 + (v2_normalized - v1) *\
      step / (num_steps - 1)
    interpolated_norm = tf.norm(interpolated)
    interpolated_normalized =\
      interpolated * (v1_norm / interpolated_norm)
    vectors.append(interpolated_normalized)
  return tf.stack(vectors)
```

The function initiates latent space interpolation between two randomly initialized vectors. It interpolates between vectors that are nonzero and don't both lie on a line going through the origin and returns the normalized interpolated vectors to the calling environment. **Image interpolation** is the process of generating in-between images from a sequence of images.

The function begins by creating two Euclidean normed vectors v1 and v2. It then normalizes v2 to have the same norm as v1. It continues by interpolating between the two vectors on the hypersphere (or latent space) to produce a set of vectors based on the number of interpolation steps.

A **hypersphere** is a four-dimensional analog of a sphere. Although a sphere exists in 3D space, its surface is two-dimensional. Similarly, a hypersphere has a three-dimensional surface that curves into 4D space. Our universe could be the hypersurface of a hypersphere.

Load the Pre-trained Model

Set a global random seed to maintain reproducibility:

```
tf.random.set_seed(7)
```

Set the seed value. Use any number that you wish. But use the same number on all experiments for reproducibility.

Load the pre-trained Progressive GAN (progan-128):

```
hub_model = hub.load(
    'https://tfhub.dev/google/progan-128/1')\
    .signatures['default']
```

Get output shapes:

```
hub_model.output_shapes
```

The Progressive GAN (progan-128) is trained on CelebA using image size of 128 × 128 × 3 pixel images.

Get dimensions for the latent space:

```
hub_model.structured_input_signature
```

The progan-128 model maps from a 512-dimensional latent space to images. During training, latent space vectors are sampled from a normal distribution.

Generate and Display an Image

A *new* image can be generated from a random point in the latent space. Begin by creating a random normal vector in the latent space. Continue by feeding the vector to progan-128. The pre-trained model identifies the closest vector in the latent space to the random vector and generates an image from the closest vector. Algorithmically, progan-128 identifies the closest latent vector by minimizing the overall distance between the real and generated distributions.

Create a function to find the closest vector in the latent space:

```
def get_module_space_image():
  vector = tf.random.normal([1, latent_dim])
  image = hub_model(vector)['default'][0]
  return image
```

The function creates a random normal vector between 1 and 512 (the size of our latent space). It then uses progan-128 to generate an image from the closest latent vector to the random latent vector we created.

Display a generated image:

```
generated_image = get_module_space_image()
display_image(generated_image)
```

Not bad! The pre-trained model generates a relatively realistic image from a vector drawn from the latent space.

Create a Function to Generate Multiple Images

The function creates two random vectors, interpolates the space between them in the latent space, and uses progan-128 to generate images as shown in Listing 11-3.

Listing 11-3. Function to Generate Interpolated Images

```
def interpolate_between_vectors(steps):
  v1 = tf.random.normal([latent_dim])
  v2 = tf.random.normal([latent_dim])
  vectors = interpolate_hypersphere(v1, v2, steps)
  interpolated_images = hub_model(vectors)['default']
  return interpolated_images
```

The function creates two random normal variables from the 512-dimensional latent space. It then creates a tensor with *n* steps of interpolation between v1 and v2. It ends by using progan-128 to generate *n* images based on the tensor. Since the tensor consists of n steps of interpolation, generated images are along a line of interpolation bounded by the closest vectors in the latent space between v1 and v2.

Create an Animation

Create an animation based on n generated images:

```
interpolation_steps = 100
interpolated_images = interpolate_between_vectors(
    interpolation_steps)
animate(interpolated_images)
```

Pretty amazing! With the help of progan-128, we create an animation based on interpolation between two vectors in the latent space closest to the two random vectors we created. The number of steps makes a big difference in the animation process! The higher the number of steps, the more interpolated images are created between the vectors in the latent space. So experiment with the number of steps to see what happens. But don't use too big of an n size because you may run out of memory!

Tip Keep interpolation steps relatively low to avoid a memory crash! Although Colab is a cloud service, the free version is pretty limited in the amount of allocated RAM to a notebook.

Display Interpolated Image Vectors

Display interpolated image vectors to deconstruct the image generation process.

Get the number of interpolated images in the latent space bounded by random vectors v1 and v2:

```
num_imgs = len(interpolated_images)
num_imgs
```

We set interpolation steps to 100, so we have 100 interpolated images.

Display the initial interpolated image:

```
show_image(interpolated_images[0])
```

Display the final interpolated image:

```
show_image(interpolated_images[num_imgs - 1])
```

So the animation begins with the initial image and morphs into the final image because interpolated vectors are along a line in the latent space bounded by random vectors v1 and v2.

Create a function to display generated image vectors from the latent space as shown in Listing 11-4.

Listing 11-4. Function to Display Generated Image

```
def generated_images(images, cols, rows):
  columns, rows = cols, rows
  ax = []
  fig = plt.figure(figsize=(20, 20))
  for i in range(columns*rows):
    img = images[i].numpy()
    ax.append(fig.add_subplot(rows, columns, i+1))
    plt.imshow(img)
    plt.axis('off')
```

Display interpolated images:

```
generated_images(interpolated_images, 10, 10)
```

It's fascinating to see how the model creates a series of images that morph from the initial to final one based on interpolated vectors from the latent space! So a learning model can create a series of vectors on a linear path between two points in the latent space as we demonstrated with our animation.

Display Multiple Images from a Single Vector

Create a function to return a latent vector and image:

```
def get_vector():
  vector = tf.random.normal([1, latent_dim])
  image = hub_model(vector)['default'][0]
  return vector, image
```

The function returns a random normal latent vector and an image generated by progan-128 from the random latent vector. Each time the function is executed, images are different because each latent vector is generated from a random vector, fed into progan-128, and generated.

Display:

```
for _ in range(2):
  latent_vector, image = get_vector()
  print (latent_vector[0][0:3])
  show_image(image)
```

For each loop cycle, a slice from each latent vector and its corresponding image are displayed. The two images are dissimilar and have no relationship because we create a latent vector close to a *random vector* for each loop cycle. But there is no path between the vectors because we didn't create a series of latent vectors on a linear path bounded by two random vectors! We only interpolated a single vector from the latent space for each loop cycle.

Display several images as shown in Listing 11-5.

Listing 11-5. Display Several Images Based on a Single Random Vector

```
rows, cols = 2, 2
plt.figure(figsize=(10, 10))
for i in range(rows*cols):
  plt.subplot(rows, cols, i + 1)
  plt.imshow(get_module_space_image())
  plt.axis('off')
```

Again, images are dissimilar and have no relationship because the function creates a latent vector close to a random latent vector each time it is invoked!

Create a Target Latent Vector from an Uploaded Image

As we've already demonstrated, we can create a random normal latent vector and generate a new image from it. Alternatively, we can create a vector from an uploaded image vector and generate a new image from it.

Import the requisite library:

```
from google.colab import files
```

Create a function to get an uploaded image from your local drive:

```
def upload_image():
  uploaded = files.upload()
  image = imageio.imread(uploaded[list(uploaded.keys())[0]])
  return transform.resize(image, [128, 128])
```

The function uses *files.upload()* from the *google.colab* library to enable file uploads to the notebook. It uses *imread()* from the *imageio* library to read the image from a local drive. The function returns a resized image in the form expected by progan-128 using *transform.resize()* from the *skimage* library.

Note Be sure to have an image on your local drive to upload. As a convenience, we recommend using one of the images included on the website for this chapter. Just copy the image to your local drive.

Get an image from your local drive:

```
local_image = upload_image()
```

Click *Choose Files* to select the image you want to load from your local drive.

Display the image:

```
display_image(local_image)
```

Create a generated image based on the local image vector:

```
vector = tf.dtypes.cast(local_image, tf.float32)
generated_image = hub_model(vector)['default'][0]
display_image(generated_image)
```

Instead of creating a random normal latent vector to generate an image, create a vector from an uploaded image to generate an image. We do need to convert the local image tensor to a float tensor. Continue by generating an image from the float tensor with the help of progan-128. End by displaying the image. The progan-128 model generates a new image based on what it learned from CelebA. So a generated image resembles a human face regardless of what image is uploaded because progan-128 learned on human faces! Although deep learning is amazing, it is still limited to what it is trained upon (at least for the time being).

Custom Loop Learning Experiment

The objective of a supervised learning experiment is to predict the correct label for newly presented input data. The model learns how to connect input features to a predicted outcome in a very precise manner. Although GANs are unsupervised experiments, we can create a *pseudo-supervised* experiment by creating a series of latent vectors on a linear path bounded by two randomly generated vectors. We identify one of the generated vectors as the feature vector and the other as the predicted (or target) image. A **latent vector** is one that is drawn from the latent spacc that cannot be accessed or manipulated during training. The idea is similar to how feedforward neural networks cannot manipulate values output by hidden layers.

We train the model by defining a loss function between the feature vector and target image and use gradient descent on the feature vector to find variable values that minimize the loss. Training begins by defining a loss function between the feature image and target image. A custom training loop enables a gradient descent algorithm and the loss function to find variable values that minimize the loss between the feature vector and target.

Create the Feature Vector

Generate a seed for reproducibility and create a feature vector:

```
seed_value = 777
tf.random.set_seed(seed_value)
feature_vector = tf.random.normal([1, latent_dim])
```

Change the seed to generate a different image. Be careful to use the same seed value to maintain reproducibility between experiments.

Note We set a different random seed for this experiment. We could have kept the global random seed that we set in the first experiment. But we decided to separate the seeds for the two experiments.

The feature vector can be thought of as the initial vector because the model learns through gradient steps how to proceed from an initial latent vector to the final target image.

Verify that the feature vector was drawn from the latent space:

```
feature_vector.shape
```

The feature vector is drawn from the 512-dimensional latent space as indicated by shape 1 × 512. So the feature vector is a one-dimensional vector with 512 elements.

Display an Image from the Feature Vector

Display an image from the feature vector using progan-128:

```
display_image(hub_model(feature_vector)['default'][0])
```

Create the Target Image

Create the target image from the latent space:

```
target_image = get_module_space_image()
```

Verify its shape:

```
target_image.shape
```

The shape of the target tensor is 128 × 128 × 3, which is the shape expected by progan-128.

Display the target image:

```
display_image(target_image)
```

Create a Function to Find the Closest Latent Vector

The function defines a loss function between the feature vector and target image. It uses gradient descent to find variable values that minimize loss as shown in Listing 11-6.

Listing 11-6. Custom Training Loop Function

```
def find_closest_latent_vector(
    initial_vector, target_image,
    num_optimization_steps,
    steps_per_image, loss_alg):
  images = []
  losses = []
  vector = tf.Variable(initial_vector)
  optimizer = tf.optimizers.Adam(learning_rate=0.01)
  loss_fn = loss_alg
  for step in range(num_optimization_steps):
    if (step % 100)==0:
      print()
    print('.', end='')
    with tf.GradientTape() as tape:
      image = hub_model(vector.read_value())['default'][0]
      if (step % steps_per_image) == 0:
        images.append(image.numpy())
      target_image_difference = loss_fn(
          image, target_image[:,:,:3])
      regularizer = tf.abs(tf.norm(vector) - np.sqrt(latent_dim))
      loss = target_image_difference + regularizer
      losses.append(loss.numpy())
    grads = tape.gradient(loss, [vector])
    optimizer.apply_gradients(zip(grads, [vector]))
  return images, losses
```

The function accepts the feature vector, number of steps, steps per image, and loss algorithm. It then initializes variables including the optimizer and loss function. It also converts the feature vector to a tf.Variable. A *tf.Variable* is a tensor whose value can be changed during training.

The function continues with a custom training loop. For custom loop learning, TensorFlow provides the *tf.GradientTape* API for automatic differentiation. TensorFlow records relevant operations executed inside the context of a tf.GradientTape onto a virtual tape. TensorFlow then uses the virtual tape to compute the gradients of a recorded computation using reverse-mode differentiation.

Automatic differentiation (AD) is a set of techniques to numerically evaluate the derivative of a function specified by a computer program. We use AD to compute the gradient of a computation with respect to some inputs (e.g., tf.Variables). AD implements two distinct modes. *Forward-mode differentiation* traverses the chain rule from inside out. *Reverse-mode differentiation* traverses the chain rule from outside to inside. Automatic differentiation is useful for implementing ML algorithms such as back-propagation for training neural networks.

For an excellent resource on automatic differentiation, peruse

https://rufflewind.com/2016-12-30/reverse-mode-automatic-differentiation

To differentiate automatically, TensorFlow needs to remember what operations happen in what order during the forward pass. Then, during the backward pass, TensorFlow traverses this list of operations in *reverse order* to compute gradients.

The gradient tape uses the pre-trained model to generate an image, finds the space between the feature image and target image, uses regularization to get more realistic images, calculates loss, applies the gradient descent algorithm, and optimizes the gradients. Once training is completed, the function returns an array of generated images and an array of losses calculated during training.

Since latent vectors are sampled from a normal distribution, we can get more realistic images if we regularize the length of the latent vector to the average length of all vectors from this distribution. We implement regularization with the *regularizer* variable in the gradient tape section of the function.

Create the Loss Function

Create the loss function algorithm:

```
reduction = tf.keras.losses.Reduction.SUM
mae_loss_algorithm = tf.losses.MeanAbsoluteError(reduction)
```

Use Mean Absolute Error (MAE) reduction for the loss function algorithm. **MAE** measures the difference between two continuous variables.

Generally, the *tf.losses.MeanAbsoluteError* API computes the mean of absolute differences between labels and predictions. In our experiment, it computes the mean absolute difference between latent vectors and the actual target.

Train the Model

Clear previous model sessions:

```
tf.keras.backend.clear_session()
```

Tip When running multiple training sessions in a notebook, it is a good idea to clear previous model sessions prior to training a model with the tf.keras.backend.clear_session API.

Invoke the training function:

```
num_optimization_steps = 200
steps_per_image = 5
mae_images, mae_loss = find_closest_latent_vector(
    feature_vector, target_image, num_optimization_steps,
    steps_per_image, mae_loss_algorithm)
```

Tweak optimization steps and steps per image to see the impact on the visualization.

Training Loss

Plot loss:

```
plt.plot(mae_loss)
fig = plt.ylim([0, max(plt.ylim())])
```

Calculate the final loss with MAE reduction:

```
MAE_loss = mae_loss[num_optimization_steps - 1]
MAE_loss
```

Animate

Create an animation from the images generated from training:

```
animate(np.stack(mae_images))
```

Compare Learned Images to the Target

Get number of generated images:

```
len(mae_images)
```

Get image type:

```
type(mae_images[0])
```

Create a function to display learned images as shown in Listing 11-7.

Listing 11-7. Function to Display Generated Images

```
def closest_latent_images(faces, rows, cols):
  fig = plt.figure(1, (20., 40.))
  for i in range(40):
    plt.subplot(10, 4, i+1)
    plt.imshow(faces[i])
    plt.axis('off')
```

Display the closest latent images between the feature and target:

```
closest_latent_images(mae_images, 10, 4)
```

During each step of the training loop, the function generates a new image by leveraging the pre-trained weights from progan-128. It then computes the loss between the newly generated image and the target. Gradually, through gradient descent and loss minimization techniques, images become more and more similar to the target. It's magic!

Contrast the first generated image against the target:

```
display_image(np.concatenate(
    [mae_images[0], target_image], axis=1))
```

Show how well the model performed:

```
display_image(np.concatenate(
    [mae_images[-1], target_image], axis=1))
```

Grab 'mae_images[-1]' to display the final generated image.

To get larger images, use the other display function:

```
show_image(np.concatenate(
    [mae_images[-1], target_image], axis=1))
```

Not bad at all!

Try a Different Loss Function Algorithm

Use a MSE instead of MAE:

```
reduction = tf.keras.losses.Reduction.SUM
mse_loss_algorithm = tf.losses.MeanSquaredError(reduction)
```

Mean Squared Error (MSE) measures the average of the squares of the errors (average squared difference) between the estimated values and the target. The *tf.losses.MeanSquaredError* API computes the mean of squares of errors between labels and predictions.

Clear previous model sessions:

```
tf.keras.backend.clear_session()
```

Train the model with MSE reduction:

```
num_optimization_steps = 200
steps_per_image = 5
mse_images, mse_loss = find_closest_latent_vector(
    feature_vector, target_image, num_optimization_steps,
    steps_per_image, mse_loss_algorithm)
```

Plot loss:

```
plt.plot(mse_loss)
fig = plt.ylim([0, max(plt.ylim())])
```

Calculate the final loss with MSE reduction:

```
MSE_loss = mse_loss[num_optimization_steps - 1]
MSE_loss
```

MSE loss is much lower.

Display the animation:

```
animate(np.stack(mse_images))
```

Compare the final generated image with the target image:

```
display_image(np.concatenate(
    [mse_images[-1], target_image], axis=1))
```

Compare to MAE reduction:

```
display_image(np.concatenate(
    [mae_images[-1], target_image], axis=1))
```

Create a Target from an Uploaded Image

Instead of creating a target image with progan-128 from a latent vector, create a target image with progan-128 from an uploaded image vector.

Generate a seed and create an initial feature vector from the latent space:

```
seed_value = 0
tf.random.set_seed(seed_value)
feature_vector = tf.random.normal([1, latent_dim])
```

Notice that we use a different seed value. Of course, you can use any seed value you wish. But always use the same seed value across experiments for reproducibility.

Grab an image from a local drive and display:

```
uploaded_image = upload_image()
display_image(uploaded_image)
```

Convert the uploaded image to float32 for TensorFlow consumption:

```
uploaded_vector = tf.dtypes.cast(uploaded_image, tf.float32)
display_image(uploaded_vector)
```

Note The uploaded tensor is *not* the target image because it was not drawn from the latent space. The target image is created from progan-128 based on the uploaded vector! As a convenience, we recommend using one of the images included on the website for this chapter. Just copy the image to your local drive.

Use progan-128 to generate the target image from the vector we just created from the uploaded image:

```
uploaded_target = hub_model(uploaded_vector)['default'][0]
display_image(uploaded_target)
```

So the uploaded image is not the target image. It is just a vector (once we convert the image to a float tensor) that progan-128 uses to create a target.

Create a loss algorithm with MSE reduction:

```
reduction = tf.keras.losses.Reduction.SUM
loss_algorithm = tf.losses.MeanSquaredError(reduction)
```

We use MSE reduction, but you can substitute MAE reduction if you wish.

Clear previous model sessions:

```
tf.keras.backend.clear_session()
```

Train:

```
num_optimization_steps = 300
steps_per_image = 5
mse_images, mse_loss = find_closest_latent_vector(
    feature_vector, uploaded_target, num_optimization_steps,
    steps_per_image, loss_algorithm)
```

We trained the model on more optimization steps to generate a better facsimile of the target image.

Calculate the final loss:

```
MSE_loss = mse_loss[num_optimization_steps - 1]
MSE_loss
```

Animate:

```
animate(np.stack(mse_images))
```

Compare the final generated image to the target:

```
display_image(np.concatenate(
    [mse_images[-1], uploaded_target], axis=1))
```

Not bad. We can increase the number of optimization steps to generate a more realistic image. But be careful because setting the number of steps too high might compromise available RAM.

Create a Target from a Google Drive Image

Instead of grabbing an image from a local drive, grab it from Google Drive. We create a new feature vector, but you can use the one we created for the uploaded image exercise if you wish.

Generate a seed and create an initial feature vector from the latent space:

```
seed_value = 0
tf.random.set_seed(seed_value)
feature_vector = tf.random.normal([1, latent_dim])
```

Mount Google Drive:

```
from google.colab import drive
drive.mount('/content/gdrive')
```

Click the URL, choose a Gmail account, copy the authorization code, paste it into the textbox, and click the *Enter* button on your keypad.

Get and display the image:

```
from PIL import Image

p1 = 'gdrive/My Drive/Colab Notebooks/'
p2 = 'images/honest_abe.jpeg'
path = p1 + p2
img_path = path
gdrive_image = Image.open(img_path)
plt.axis('off')
_ = plt.imshow(gdrive_image)
```

Create a path to the image on Google Drive. Open the image with the path and display. Be sure that the image is in the *Colab Notebooks* directory!

Note Any image (or file) that you wish to load into a Colab notebook must be in the "Colab Notebooks" directory (or a subdirectory inside the "Colab Notebooks" directory) on your Google Drive. The "Colab Notebooks" directory is automatically created the first time you save a Colab notebook. All Colab notebooks are saved to the "Colab Notebooks" directory. As a convenience, we recommend using one of the images included on the website for this chapter. Just copy the image to your Google Drive into the "Colabs Notebook" directory.

Create a function to convert the Google Drive image to a TensorFlow consumable tensor:

```
def reformat(img, size):
  img = tf.keras.preprocessing.image.img_to_array(img) / 255.
  img = tf.image.resize(img, size)
  return img
```

The Keras utility converts a PIL image instance to a NumPy array and resizes it to the size expected by progan-128, which is 128 × 128 × 3.

Invoke the function:

```
img_size = (128, 128)
gdrive_vector = reformat(gdrive_image, img_size)
gdrive_vector.shape
```

Display the image:

```
display_image(gdrive_vector)
```

Note The uploaded tensor is *not* the target image because it was not drawn from the latent space. The target image is created from progan-128 based on the uploaded tensor!

Use progan-128 to generate the target image from the vector we just created from the Google Drive image:

```
gdrive_target = hub_model(gdrive_vector)['default'][0]
display_image(gdrive_target)
```

Again, the image loaded from Google Drive is not the target image. It is just a vector (once we convert the image to a NumPy array) based on the Google Drive image that progan-128 uses to create a target image.

Create a loss algorithm with MSE reduction:

```
reduction = tf.keras.losses.Reduction.SUM
loss_algorithm = tf.losses.MeanSquaredError(reduction)
```

We use MSE reduction, but MAE reduction should also work.

Clear previous model sessions:

```
tf.keras.backend.clear_session()
```

Train:

```
num_optimization_steps = 300
steps_per_image = 5
mse_images, mse_loss = find_closest_latent_vector(
    feature_vector, gdrive_target, num_optimization_steps,
    steps_per_image, loss_algorithm)
```

We trained the model on more optimization steps to generate a better facsimile of the target image.

Calculate the final loss:

```
MSE_loss = mse_loss[num_optimization_steps - 1]
MSE_loss
```

Animate:

```
animate(np.stack(mse_images))
```

Compare the final generated image to the target:

```
display_image(np.concatenate(
    [mse_images[-1], gdrive_target], axis=1))
```

Create a Target from a Wikimedia Commons Image

Wikimedia Commons is an online repository of free-use images, sounds, other media, and JSON files. It is a project of the Wikimedia Foundation. Wikimedia Commons only accepts free content such as images and other media files not subject to copyright restrictions that would prevent them from being used by anyone anytime and for any purpose.

Generate a seed and create a feature vector:

```
seed_value = 0
tf.random.set_seed(seed_value)
feature_vector = tf.random.normal([1, latent_dim])
```

Grab an image:

```
p1 = 'http://upload.wikimedia.org/wikipedia/commons/'
p2 = 'd/de/Wikipedia_Logo_1.0.png'
URL = p1 + p2
im = imageio.imread(URL)
im.shape
```

Import the requisite library for converting an image tensor to a NumPy array:

```
from keras.preprocessing.image import img_to_array
```

Convert the image to a NumPy array and display:

```
img_array = img_to_array(im)
print(img_array.dtype)
print(img_array.shape)
plt.imshow(tf.squeeze(img_array))
fig = plt.axis('off')
```

The Keras utility converts a PIL image instance to a NumPy array so we can display the image.

Resize the image for progan-128 consumption:

```
wiki_vector = tf.image.resize(img_array, (128, 128))
plt.imshow(tf.squeeze(wiki_vector))
fig = plt.axis('off')
```

Resize the image to the expected size of progan-128 and display it.

Create the target:

```
wiki_target = hub_model(wiki_vector)['default'][0]
display_image(wiki_target)
```

Create a loss algorithm with MSE reduction:

```
reduction = tf.keras.losses.Reduction.SUM
loss_algorithm = tf.losses.MeanSquaredError(reduction)
```

Clear previous model sessions:

```
tf.keras.backend.clear_session()
```

Train:

```
num_optimization_steps = 300
steps_per_image = 5
mse_images, mse_loss = find_closest_latent_vector(
    feature_vector, wiki_target, num_optimization_steps,
    steps_per_image, loss_algorithm)
```

Calculate the final loss:

```
MSE_loss = mse_loss[num_optimization_steps - 1]
MSE_loss
```

Animate:

```
animate(np.stack(mse_images))
```

Compare:

```
display_image(np.concatenate(
    [mse_images[-1], wiki_target], axis=1))
```

Not bad.

Although we generate pretty realistic images and facsimiles of a target image, much more robust models and training algorithms are needed to consistently generate images that are indistinguishable from actual ones. And, of course, more computing resources are needed!

Latent Vectors and Image Arrays

The progan-128 module generates a new image from either a latent vector of size (1, 512) or float vector of size 128 × 128 × 3. A latent vector accepted by progan-128 is a one-dimensional vector of size 512. A float vector accepted by progan-128 is a 128 × 128 × 3-pixel vector.

The progan-128 module is an image generator based on the TensorFlow re-implementation of Progressive GANs. It maps from a 512-dimensional latent space to images. During training, latent space vectors are sampled from a normal distribution.

According to the documentation, progan-128 takes an input tensor with datatype float32 and shape (?, 512). The input tensor to progan-128 represents a batch of latent vectors. The output from progan-128 is a float tensor with shape (?, 128, 128, 3), which represents a batch of RGB images. We can also generate a new image from an image array, which is not included in the documentation.

To view progan-128 documentation, peruse

https://tfhub.dev/google/progan-128/1

Generate a New Image from a Latent Vector

Create a random normal vector from the latent space:

```
random_normal_latent_vector = tf.random.normal([1, latent_dim])
random_normal_latent_vector.shape
```

The tf.random.normal API outputs random values from a normal distribution. So the new vector consists of 512 randomly drawn values from a normal distribution based on our latent space.

Convert the tensor to NumPy to enable inspection:

```
rnlv = random_normal_latent_vector.numpy()
len(rnlv[0])
```

Inspect some elements from the NumPy array:

```
for i, element in enumerate(rnlv[0]):
  if i < 5:
    print (element)
  else: break
```

Each element in the new vector represents a latent dimension (or latent variable) that cannot be directly observed, but can be assumed to exist. Since latent dimensions exist, they can be used to explain patterns of variation in observed variables. In our experiment, observed variables represent CelebA images. So we can feed progan-128 the new vector to generate a new image.

Create a float output tensor from the latent space with progan-128:

```
float_output_tensor = hub_model(
    random_normal_latent_vector)['default'][0]
float_output_tensor.shape
```

Display the float output tensor as an image:

```
display_image(generated_image)
```

So progan-128 generates a 128 × 128 × 3 image from a latent vector.

Generate a New Image from an Image Vector

Get an image from Google Drive:

```
p1 = 'gdrive/My Drive/Colab Notebooks/'
p2 = 'images/honest_abe.jpeg'
path = p1 + p2
img_path = path
abe_image = Image.open(img_path)
plt.axis('off')
_ = plt.imshow(abe_image)
```

Convert the JPEG image to an image vector of the appropriate datatype and size:

```
img_size = (128, 128)
abe_vector = reformat(abe_image, img_size)
abe_vector.shape
```

Display a slice from the vector:

```
abe_vector[0][0].numpy()
```

The vector is definitely not from the latent space!

Generate a new image from the vector we just created:

```
image_from_abe_vector = hub_model(abe_vector)['default'][0]
display_image(image_from_abe_vector)
```

So progan-128 generates a 128 × 128 × 3 image from a feature image vector not drawn from the latent space!

Summary

We demonstrated two Progressive Growing GAN experiments in a detailed and step-by-step manner. We also verified that progan-128 can also generate a realistic image from a feature vector not drawn from the latent space.

CHAPTER 12

Fast Style Transfer

Neural style transfer (NST) is a computer vision technique that takes two images – a content image and a style reference image – and blends them together so that the resulting output image retains the core elements of the content image but appears to be painted in the style of the style reference image. The output image from a NST network is called a pastiche. A **pastiche** is a work of visual art, literature, theater or music that imitates the style (or character) of the work of one or more other artists. Unlike a parody, a pastiche celebrates rather than mocks the work it imitates.

In many applications of NST, the content image is a photograph, and the style reference image is a painting. But this is not a requirement. The wonder of NST is that we can take an artwork by a famous painter and blend it so the output image looks like the content image painted in the reference image style. So we can train NST networks to automatically create new artistic renderings!

Note The terms network and model are interchangeable in the data science vernacular (at least when working with neural network models).

The technique is implemented by optimizing the output image to match the content statistics of the content image and the style statistics of the style reference image. Statistics are extracted from the images using a convolutional network. The idea is to train the network to match a base input image with the content and style reference images. NST transforms the base input image by minimizing the content and style reference distances (or losses) with back-propagation to create an image that matches the content of the content image and the style of the style reference image.

To extract the content and style representations into a pastiche, the neural net includes intermediate layers. Intermediate layers represent feature maps that become increasingly higher-ordered as we go deeper into the network to define the representation of content and style from the respective images. During training, the

D. Paper, *State-of-the-Art Deep Learning Models in TensorFlow*, https://doi.org/10.1007/978-1-4842-7341-8_12

network tries to match the base input image to the corresponding style and content target representations at the intermediate layers.

For a network to perform image classification, it must understand the image. So it takes a raw image as input pixels and builds an internal representation through transformations that turn raw image pixels into a complex understanding of the features present within the image. The *intermediate layers* of the network perform the transformations that allow the model to extract meaningful features from the raw input pixels of the images. So the intermediate layers are able to describe the content and style of input images with precision.

Despite amazing results, NST operates at a slow pace because it treats the task as an optimization problem with hundreds or even thousands of iterations to perform style transfer on just a single image! To tackle this inefficiency, deep learning researchers developed a technique referred to as fast (neural) style transfer. **Fast style transfer** uses deep neural networks, but trains a standalone model to transform an image in a *single* feedforward pass! So trained fast style transfer models can stylize any image with just one iteration (or epoch) through the network instead of hundreds or thousands.

Why Is Style Transfer Important?

Not everyone is a born artist. But with recent advances in technologies like style transfer, almost anyone can enjoy the pleasure that comes along with creating and sharing an artistic masterpiece.

With the transformative power of style transfer, artists can easily lend their creative aesthetic to others. People without inherent artistic ability can create new and innovative representations of artistic styles to live alongside original masterpieces. So style transfer empowers people to cultivate their own creativity! Creative transformations may lead to new artistic representations that may never have otherwise been created.

Arbitrary Neural Artistic Stylization

Although fast style transfer networks are fast, they are limited to a preselected handful of styles because a separate neural network must be trained for each style image. *Arbitrary Neural Artistic Stylization* (ANAS) mitigates this limitation by using a style network and transformer network. The *style network* learns how to break down an image into a 100-dimensional vector (or style vector) that represents its style. The *transformer network* learns how to produce the final stylized image from the style vector and original content image.

Note Style vectors are also referred to as style bottleneck vectors.

ANAS networks are the latest incarnation of fast style networks. ANAS are considered better than NST and fast style transfer because they enable arbitrary style transfer in real time. So they are faster than NST and more flexible than fast style transfer because they automatically adapt to arbitrary new styles.

ANAS adds an *adaptive instance normalization* (AdaIN) layer that aligns the mean and variance of the content features with those of the style features. ANAS achieves speeds comparable to the fastest existing approaches without the restriction of a predefined set of styles. ANAS also allows flexible user controls such as content-style trade-off, style interpolation, and color and spatial controls with only a single feedforward neural network pass!

Notebooks for chapters are located at the following URL:

https://github.com/paperd/deep-learning-models

Since ANAS is the fastest and best implementation of NST, we demonstrate it with an end-to-end code experiment. We include a second ANAS experiment using a pre-trained transfer model from the TensorFlow Lite module.

TensorFlow Lite is a set of tools to help developers run TensorFlow models on mobile, embedded, and IoT (Internet of Things) devices. It enables on-device ML inference with low latency and a small binary size. An **IoT** device is one that has a sensor attached to it and can transmit data from one object to another or to people with the help of the Internet. *Low latency* describes a computer network that is optimized to process a very high volume of data messages with minimal delay (or latency). Such networks are designed to support operations that require near-real-time access to rapidly changing data.

Begin setting up the Colab ecosystem by importing the main TensorFlow library and instantiating the GPU.

Import the TensorFlow Library

Import the library and alias it as **tf**:

```
import tensorflow as tf
```

Aliasing the TensorFlow library as tf is common practice.

GPU Hardware Accelerator

As a convenience, we include the steps to enable the GPU in a Colab notebook:

1. Click *Runtime* in the top-left menu.
2. Click *Change runtime type* from the drop-down menu.
3. Choose *GPU* from the *Hardware accelerator* drop-down menu.
4. Click *Save*.

Verify that the GPU is active:

```
tf.__version__, tf.test.gpu_device_name()
```

If '/device:GPU:0' is displayed, the GPU is active. If '..' is displayed, the regular CPU is active.

Note If you get the error **NAME 'TF' IS NOT DEFINED**, re-execute the code to import the TensorFlow library!

ANAS Experiment

We use the *arbitrary-image-stylization-v1-256* network for the experiment. The network is a pre-trained model for fast arbitrary image style transfer. The network doesn't require that images be resized, but prefers that style reference images are about 256 pixels because it is trained on 256 × 256 pixel images. But content images can be any size.

Note We found that style reference images sized larger or smaller than 256 × 256 pixels don't provide very compelling pastiches. So we highly recommend resizing style reference images to the preferred size.

Import Requisite Libraries

Import:

```
from matplotlib import gridspec
import matplotlib.pylab as plt
import numpy as np
import tensorflow_hub as hub
from PIL import Image
```

The *gridspec* module specifies the geometry of the grid where a subplot is placed. Setting the number of rows and number of columns of the grid is a requirement. *TensorFlow Hub* is a repository of pre-trained machine learning models that can be tuned and deployed anywhere with just a few lines of code. The *Image* module is used to represent a Python Imaging Library (PIL) image. The *PIL* module provides a number of factory functions including those to load images from files and create new images.

Get Images from Google Drive

We get images for this experiment from Google Drive. Other options include (but are not limited to) uploading images from a local drive or downloading images from Wikipedia Commons.

Note Be sure that all images are in the *Colab Notebooks* directory on Google Drive.

Mount Google Drive:

```
from google.colab import drive
drive.mount('/content/gdrive')
```

Click the URL. Choose a Google Gmail account. Click the *Allow* button. Copy and paste the authorization code into the textbox and press the *Enter* key on your keypad.

Load and display the style reference image:

```
img_path = 'gdrive/My Drive/Colab Notebooks/images/serene.jpeg'
style = Image.open(img_path)
plt.axis('off')
_ = plt.imshow(style)
```

Images are available through the Apress website or from my GitHub.

Get the image type:

```
type(style)
```

The image is a PIL image.

Get the shape of the style reference image:

```
w, h = style.size
w, h
```

The *size* method returns the shape of a PIL image.

Load and display the content image:

```
img_path = 'gdrive/My Drive/Colab Notebooks/images/'\
  'humming_bird.jpeg'
content  = Image.open(img_path)
plt.axis('off')
_ = plt.imshow(content)
```

Get the image type:

```
type(content)
```

Get the shape of the content image:

```
w, h = content.size
w, h
```

Preprocess Images

Since both images are PIL images, we convert them to tensors so they are consumable by the pre-trained model. The recommended size for the style reference image is 256 × 256 because this is the size expected by the pre-trained style transfer network we use for this experiment. The content image can be any size.

Convert the style reference image to a NumPy array and scale it:

```
style_array = tf.keras.preprocessing.image.img_to_array(
    style) / 255.
style_array.shape
```

Resize the style reference image to what is expected by the style transfer network:

```
style_img = tf.image.resize(style_array, (256, 256))
style_img.shape
```

Voilà. The style reference image is ready to be consumed by the style transfer network.

Convert the content image to a NumPy array and scale it:

```
content_img = tf.keras.preprocessing.image.img_to_array(
    content) / 255.
content_img.shape
```

Since the content image can be any size, it is now ready to be consumed by the style transfer network.

Display Processed Images

Create a display function:

```
def display_one(img):
  plt.imshow(img)
  plt.axis('off')
  plt.show()
```

Display the processed style reference image:

```
display_one(style_img)
```

Display the processed content image:

```
display_one(content_img)
```

Prepare Image Batches

Although the processed content and style reference images are consumable by the style transfer network, it expects each image as a separate batch. So each batch must be a 4D tensor with shape *[batch_size, image_height, image_width, 3]*. Since content and style reference images are currently 3D tensors with shapes *[image_height, image_width, 3]*, we add a batch dimension of "1" to satisfy the requirement. The network also requires images as TensorFlow tensors. So we convert them to TensorFlow tensors.

Input and output values of the images are expected to be in the range [0, 1]. We already satisfied this requirement by scaling both images. Shapes of content and style reference images don't have to match. So we are fine in that regard.

Note The pastiche (output image) shape is adapted from the content image shape.

Add the batch dimension to both images:

```
style_image = np.expand_dims(style_img, axis=0)
content_image = np.expand_dims(content_img, axis=0)
style_image.shape, content_image.shape
```

Convert NumPy images to TensorFlow tensors:

```
style_tensor = tf.convert_to_tensor(style_image)
content_tensor = tf.convert_to_tensor(content_image)
```

Load the Model

As stated earlier, we use the *arbitrary-image-stylization-v1-256* network to create a pastiche. Earlier NST models are limited to a pre-selected handful of styles because a separate neural network must be trained for each style image. Arbitrary style transfer mitigates this limitation by including a style network and transformer network.

The *style network* learns how to break down an image into a 100-dimensional vector (or style bottleneck vector) that represents its style. The *transformer network* learns how to produce the final stylized image from the style bottleneck vector and original content image. Arbitrary style networks are state-of-the-art (as of this writing) because they are faster than original NST networks and more flexible than fast style transfer networks.

Load the pre-trained ANAS network:

```
p1 = 'https://tfhub.dev/google/magenta/'
p2 = 'arbitrary-image-stylization-v1-256/2'
URL = p1 + p2

hub_handle = URL
hub_module = hub.load(hub_handle)
```

Build the hub module from a pre-trained ANAS network.

Feed the Model

Demonstrate arbitrary style transfer:

```
outputs = hub_module(content_tensor, style_tensor)
pastiche = outputs[0]
pastiche.shape
```

The signature of the hub module accepts the processed content image and the processed style reference image to create a pastiche by learning how to blend the two tensors. A *signature* is the procedural syntax required to train a model.

Training is fast with the GPU! But training the hub module signature with a CPU takes some time.

Explore the Pastiche

Explore image shapes:

```
pastiche_numpy = tf.squeeze(pastiche).numpy()
pastiche_numpy.shape, content_img.shape
```

Convert the pastiche to NumPy for easy exploration. The shape is not exactly the same as the content image, but very close.

Explore a slice from the stylized image tensor:

```
pastiche_numpy[0][0]
```

The pastiche has the same pixel characteristics as the content and style reference images.

Extract matrix components:

```
m = pastiche_numpy
r, c, channels = m.shape[0], m.shape[1], m.shape[2]
r, c, channels
```

The pastiche array is a 3D matrix consisting of 184 rows, 280 columns, and 3 channels.

Get the number of pixels in the matrix:

```
pixels = r * c
pixels
```

Each RGB channel has 51,520 pixels. **RGB** refers to three hues of light (red, green, and blue) that can be mixed together to create different colors. Combining red, green, and blue lights is the standard method for producing color images on screens such as TVs, computer monitors, and smartphones. The *RGB color model* is an additive model because the three light beams of color are added together (light spectra wavelength by light spectra wavelength) to create the final color spectrum.

Check if RGB channel pixels are scaled:

```
red = m[m[:, :, 0] < 1, 0] < 1
green = m[m[:, :, 1] < 1, 0] < 1
blue = m[m[:, :, 2] < 1, 0] < 1
print (len(red), len(green), len(blue))
print (red, green, blue)
```

The algorithms check if the pixels in each channel are less than one. We display the length of each algorithmically modified channel to see if they contain the expected number of pixels. So far everything checks out.

Since all truth table values aren't displayed, we need one more step to verify that scaling worked as expected:

```
all(red), all(green), all(blue)
```

The *all()* function returns True if all items in an iterable are true; otherwise, it returns False. So all pixels are scaled. An **iterable** is an object that can be iterated over. A Python list is an example of an iterable.

But we can check if all matrix pixels are scaled in *one* step:

```
truth = np.where((m < 1), True, False)
truth.all()
```

Note We demonstrate the multistep process to provide a glimpse under the hood of the stylized image.

Visualize

Create a visualization function as shown in Listing 12-1.

Listing 12-1. Visualization Function

```
def show_n(images, titles=('',)):
  n = len(images)
  image_sizes = [image.shape[1] for image in images]
  w = (image_sizes[0] * 6) // 320
  plt.figure(figsize=(w  * n, w))
  gs = gridspec.GridSpec(1, n, width_ratios=image_sizes)
  for i in range(n):
    plt.subplot(gs[i])
    plt.imshow(images[i][0], aspect='equal')
    plt.axis('off')
    plt.title(titles[i] if len(titles) > i else '')
  plt.show()
```

Visualize the original content, style reference, and stylized images:

```
show_n([content_image, style_image, pastiche],
       titles=['Original content image', 'Style image',
               'Pastiche'])
```

Create a function to visualize the new stylized image:

```
def display_pastiche(img, size):
  plt.figure(figsize = size)
  plt.imshow(tf.squeeze(img))
  plt.axis('off')
  plt.show()
```

Visualize:

```
f_size = (10, 15)
display_pastiche(pastiche, f_size)
```

Image Stylization with Multiple Images

Create lists of style reference and content images. Choose one from each list to create a pastiche.

Get Images

Create a function to grab an image from Google Drive:

```
def get_image(img_path):
  return Image.open(img_path)
```

Get style images as shown in Listing 12-2.

Listing 12-2. Get Style Reference Images

```
d = 'gdrive/My Drive/Colab Notebooks/images/dali.jpg'
dali = get_image(d)
v = 'gdrive/My Drive/Colab Notebooks/images/van-gogh.jpg'
van_gogh = get_image(v)
m = 'gdrive/My Drive/Colab Notebooks/images/modern.jpg'
modern = get_image(m)
e = 'gdrive/My Drive/Colab Notebooks/images/escher.jpeg'
escher = get_image(e)
pic = 'gdrive/My Drive/Colab Notebooks/images/picasso.jpg'
picasso = get_image(pic)
```

```
p = 'gdrive/My Drive/Colab Notebooks/images/pollock.jpg'
pollock = get_image(p)
mon = 'gdrive/My Drive/Colab Notebooks/images/monet.jpg'
monet = get_image(mon)
```

Display style reference image shapes:

```
dali.size, van_gogh.size, modern.size, escher.size,\
picasso.size, pollock.size, monet.size
```

Get content images:

```
t = 'gdrive/My Drive/Colab Notebooks/images/teddy.jpeg'
teddy = get_image(t)
e = 'gdrive/My Drive/Colab Notebooks/images/einstein.jpg'
einstein = get_image(e)
g = 'gdrive/My Drive/Colab Notebooks/images/gem.jpeg'
gem = get_image(g)
```

Display content image shapes:

```
teddy.size, einstein.size, gem.size
```

Process Images

Create a preprocessing function as shown in Listing 12-3.

Listing 12-3. Preprocessing Function

```
def preprocess(img, style=True):
  img_array = tf.keras.preprocessing.image.img_to_array(
      img) / 255.
  if style:
    img_array = tf.image.resize(img_array, (256, 256))
  return\
  tf.convert_to_tensor(np.expand_dims(img_array, axis=0))
```

The function converts a PIL image into a NumPy array. It then scales, resizes, and converts the array to a TensorFlow tensor with the appropriate dimensions.

Process the style reference images as shown in Listing 12-4.

Listing 12-4. Process the Style Reference Images

```
dali_style = preprocess(dali)
van_gogh_style = preprocess(van_gogh)
modern_style = preprocess(modern)
escher_style = preprocess(escher)
picasso_style = preprocess(picasso)
pollock_style = preprocess(pollock)
monet_style = preprocess(monet)

dali_style.shape, van_gogh_style.shape, modern_style.shape,\
escher_style.shape, picasso_style.shape, pollock_style.shape,\
monet_style.shape
```

Process the content images:

```
einstein_content = preprocess(einstein, False)
teddy_content = preprocess(teddy, False)
gem_content = preprocess(gem, False)
einstein_content.shape, teddy_content.shape, gem_content.shape
```

Visualize Processed Images

Place style reference and content images in lists:

```
styles = [dali_style, van_gogh_style, modern_style,
          escher_style, picasso_style, pollock_style,
          monet_style]
contents = [einstein_content, teddy_content, gem_content]
```

Create a function to display tensors as shown in Listing 12-5.

Listing 12-5. Function to Display Tensors

```
def display_tensors(imgs, r, c):
  _, axs = plt.subplots(r, c, figsize=(12, 12))
  axs = axs.flatten()
  for img, ax in zip(imgs, axs):
```

```
    ax.imshow(tf.squeeze(img))
    ax.axis('off')
  plt.show()
```

Display style tensors:

```
rows, cols = 1, 7
display_tensors(styles, rows, cols)
```

Display content tensors:

```
rows, cols = 1, 3
display_tensors(contents, rows, cols)
```

Create Reference Dictionaries

Create a dictionary to represent style tensors:

```
style_names = {'dali' : styles[0],
               'van_gogh' : styles[1],
               'modern' : styles[2],
               'escher' : styles[3],
               'picasso' : styles[4],
               'pollock' : styles[5],
               'monet' : styles[6]}
```

Create a dictionary to represent content tensors:

```
content_names = {'einstein' : contents[0],
                 'teddy' : contents[1],
                 'gem' : contents[2]}
```

Create a Pastiche

Create a function that creates a pastiche:

```
def create(c, s):
  content_im = content_names[c]
  style_im = style_names[s]
  outputs = hub_module(content_im, style_im)
  return content_im, style_im, outputs[0]
```

Create a pastiche:

```
content, style = 'einstein', 'dali'
content_im, style_im, sim = create(content, style)

f_size = (7, 9)
display_pastiche(sim, f_size)
```

Tip Experiment with content and style reference images to create your own pastiche. And, of course, you can create a pastiche from your own images!

Create a list for visualization:

```
imgs = [content_im, style_im, sim]
```

Visualize content, style, and pastiche:

```
display_tensors(imgs, 1, 3)
```

Try one with Teddy Roosevelt:

```
content, style = 'teddy', 'picasso'
content_im, style_im, sim = create(content, style)

f_size = (8, 10)
display_pastiche(sim, f_size)
```

Visualize content, style, and pastiche:

```
imgs = [content_im, style_im, sim]
display_tensors(imgs, 1, 3)
```

Try one with gem:

```
content, style = 'gem', 'escher'
content_im, style_im, sim = create(content, style)

f_size = (8, 10)
display_pastiche(sim, f_size)
```

Visualize content, style, and pastiche:

```
imgs = [content_im, style_im, sim]
display_tensors(imgs, 1, 3)
```

TensorFlow Lite Experiment

TensorFlow Lite is an open source deep learning framework to run TensorFlow models on-device. **On-Device Portals** (ODPs) allow mobile phone users to easily browse, purchase, and use mobile content and services. An ODP platform enables operators to provide a consistent and branded on-device experience across a large portfolio of services. TensorFlow Lite provides an ODP platform to experiment with and deploy deep learning experiments on their phone.

To get started with TensorFlow Lite on your device, peruse

www.tensorflow.org/lite/examples

For an excellent TensorFlow Lite style transfer example, peruse

www.tensorflow.org/lite/examples/style_transfer/overview

But TensorFlow Lite does not have to be deployed on-device. It can be run on a PC. We run this experiment in a Colab notebook on a PC for three reasons. First, we know that our PC has enough RAM to run full-fledged TensorFlow experiments. So we won't have any issues running a TensorFlow Lite experiment! It is also nice to have an actual keyboard and a large screen. Second, we don't know what type of on-device you use (e.g., Android, iOS, etc.). Third, TensorFlow Lite modules are pre-installed in Colab!

If you want to develop on-device, we definitely recommend TensorFlow Lite because it is optimized to run on a variety of devices including mobile phones, embedded Linux devices, and microcontrollers. So TensorFlow Lite has better on-device performance and a smaller binary file size than TensorFlow.

Architecture for a Pre-trained TensorFlow Lite Model

Content images are exactly the same as what we already worked with in the first experiment. As in the first experiment, style images are transformed (or bottlenecked) into *100-dimensional style bottleneck vectors* before being fed into the style transform model. But unlike the first experiment, we manually manipulate the style image into a consumable form by the model.

The TensorFlow Lite artistic style transfer model consists of two submodels - Style Prediction Model and Style Transform Model. The *Style Prediction Model* is a pre-trained MobileNet-v2-based neural network that takes an input style reference image and transforms it into a 100-dimensional style bottleneck vector. The *Style Transform Model* is a neural network that applies a style bottleneck vector to a content image to create a pastiche.

Crop Images

Remove unwanted noise from images.

Import a requisite library for display:

```
import matplotlib as mpl

mpl.rcParams['figure.figsize'] = (12,12)
mpl.rcParams['axes.grid'] = False
```

The mpl parameters set the display size for all images for the rest of the notebook.

Get a content image:

```
content_cd = content_names['einstein']
content_cd.shape
```

The Style Transform Model expects a content image of size 384 × 384:

```
dim = [384, 384]
content_lte = tf.image.resize(content_cd, dim)
content_lte.shape
```

Centrally crop the content image:

```
content_lite = tf.image.resize_with_crop_or_pad(
    content_lte, dim[0], dim[1])
content_lite.shape
```

Cropping is one of the most basic data augmentation processes for images. The idea is to remove unwanted or irrelevant noise from the periphery of an image, change its aspect ratio or improve its overall composition.

Get a style image:

```
style_lte = style_names['modern']
style_lte.shape
```

Centrally crop the style reference image:

```
dim = [256, 256]
style_lite = tf.image.resize_with_crop_or_pad(
    style_lte, dim[0], dim[1])
style_lite.shape
```

Display Cropped Images

Create a display function as shown in Listing 12-6.

Listing 12-6. Display Function for Cropped Images

```
def imshow(image, title=None):
  if len(image.shape) > 3:
    image = tf.squeeze(image, axis=0)
  plt.axis('off')
  plt.imshow(image)
  if title:
    plt.title(title)
```

Display cropped images:

```
plt.subplot(1, 2, 1)
imshow(content_lite, 'Content Image')

plt.subplot(1, 2, 2)
imshow(style_lite, 'Style Image')
```

Stylize Images

We feed the cropped style image into the Style Prediction Model to create a style bottleneck vector. We then feed the cropped content image and newly created style bottleneck vector into the Style Transform Model to create a pastiche.

Create the Style Prediction Model

Create a function for the Style Prediction Model as shown in Listing 12-7.

Listing 12-7. Function to Create the Style Prediction Model

```
def run_style_predict(processed_style_image):
  # load the model
  interpreter = tf.lite.Interpreter(
      model_path=style_predict_path)
  # set model input
  interpreter.allocate_tensors()
  input_details = interpreter.get_input_details()
  interpreter.set_tensor(
      input_details[0]["index"], processed_style_image)
  # calculate style bottleneck
  interpreter.invoke()
  style_bottleneck = interpreter.tensor(
      interpreter.get_output_details()[0]["index"])()
  return style_bottleneck
```

The function creates a Style Prediction Model to run style prediction on the cropped style image (*style_lite*). The function accepts a processed style reference image. It then loads the style prediction path into the interpreter object (*tf.lite.Interpreter*). The interpreter is a TensorFlow Lite pre-trained MobileNet-v2-based neural network. The interpreter uses the style_predict_path to create a tflite predicted style. Input to the interpreter model is then set. The function continues by calculating the style bottleneck. The function ends by using the bottleneck calculation and the tflite predicted style to transform the style image into a 100-dimensional style bottleneck vector. Simply, the function accepts a style reference image and transforms it into a 100-dimensional style bottleneck vector.

Set the style prediction path:

```
tflite_predict = 'style_predict.tflite'
p1 = 'https://tfhub.dev/google/lite-model/magenta/'
p2 = 'arbitrary-image-stylization-v1-256/'
```

```
p3 = 'int8/prediction/1?lite-format=tflite'
URL = p1 + p2 + p3

style_predict_path = tf.keras.utils.get_file(
    tflite_predict, URL)
```

Transform the cropped style image into a 100-dimensional style bottleneck vector:

```
style_bottleneck = run_style_predict(style_lite)
print('style bottleneck vector shape:',
      style_bottleneck.shape)
```

Create the Style Transform Model

Create a function for the Style Transform Model as shown in Listing 12-8.

Listing 12-8. Function to Create the Style Transform Model

```
def run_style_transform(
    style_bottleneck, processed_content_image):
  # load the model
  interpreter = tf.lite.Interpreter(
      model_path=style_transform_path)
  # set model input
  input_details = interpreter.get_input_details()
  interpreter.allocate_tensors()
  # set content and style bottleneck
  interpreter.set_tensor(
      input_details[0]["index"], processed_content_image)
  interpreter.set_tensor(
      input_details[1]["index"], style_bottleneck)
  interpreter.invoke()
  # return the transformed content image
  return interpreter.tensor(
      interpreter.get_output_details()[0]["index"])()
```

The function creates a Style Transform Model that applies a style bottleneck vector to a content image to create a pastiche. The function accepts a style bottleneck vector and a processed content image. It then loads the style transform path into the interpreter object (tf.lite.Interpreter). The interpreter uses the style_transform_path to create a tflite transform style. Input to the interpreter model is then set. The function continues by setting the processed content image and style bottleneck vector for the interpreter. The function ends by returning the transformed content image.

Set the style transform path:

```
tflite_transform= 'style_transform.tflite'
p1 = 'https://tfhub.dev/google/lite-model/magenta/'
p2 = 'arbitrary-image-stylization-v1-256/'
p3 = 'int8/transfer/1?lite-format=tflite'
URL = p1 + p2 + p3

style_transform_path = tf.keras.utils.get_file(
    tflite_transform, URL)
```

Create the Pastiche

Stylize the content image with the style bottleneck:

```
stylized_image = run_style_transform(
    style_bottleneck, content_lite)
```

Ensure that the stylized image (or pastiche) is the size expected by the TensorFlow Lite pre-trained model:

```
pastiche = tf.image.resize(stylized_image, [384, 384])
pastiche.shape
```

Visualize the pastiche:

```
imshow(pastiche, 'Pastiche')
```

Style Blending

By blending the style of the content image into the stylized output, we make the pastiche look more like the content image.

Prepare the Content Image

Reshape the content image for consumption by the Style Prediction Model:

```
dim = [256, 256]
content_blend = tf.image.resize_with_crop_or_pad(
    content_lite, dim[0], dim[1])
content_blend.shape
```

Transform the reshaped content image into a 100-dimensional style bottleneck vector:

```
style_bottleneck_content = run_style_predict(
    content_blend)
style_bottleneck_content.shape
```

The content image is now a 100-dimensional style bottleneck vector.

Blend the Style Bottleneck Vectors

Blend the style bottleneck vector with the content-style bottleneck vector (*style_bottleneck_content*).

Define the content blending ratio (between 0 and 1):

```
content_blending_ratio = 0.5
```

The range of blending the content image into the pastiche is from 0% to 100%. To extract no style from the content image, assign 0%. Zero percent means that the blended pastiche is the same as the pastiche. To extract all of the style from the content image, assign 100%. We set the blending ratio to 50% to extract half of the style from the content image.

Get a blended style bottleneck vector from the style bottleneck and content-style bottleneck vectors:

```
style_bottleneck_blended =\
  content_blending_ratio * style_bottleneck_content +\
  (1 - content_blending_ratio) * style_bottleneck
```

Stylize the content image using the style bottleneck:

```
stylized_image_blended = run_style_transform(
    style_bottleneck_blended, content_lite)
```

Visualize the pastiche:

```
imshow(stylized_image_blended, 'Blended Stylized Image')
```

Save the Pastiche

Save the pastiche to a local drive.

Create a function to convert a tensor to a PIL image as shown in Listing 12-9.

Listing 12-9. Function to Convert a Tensor to a PIL Image

```
def tensor_to_image(tensor):
  tensor = tensor * 255
  tensor = np.array(tensor, dtype=np.uint8)
  if np.ndim(tensor) > 3:
    assert tensor.shape[0] == 1
    tensor = tensor[0]
  return Image.fromarray(tensor)
```

The function up-scales the scaled tensor and converts it to a NumPy array. It then strips the "1" dimension and returns a PIL image.

Save a file to a local drive:

```
from google.colab import files

fn = 'patiche.jpg'
tensor_to_image(stylized_image_blended).save(fn)
files.download(fn)
```

Import the *files* module from the google.colab library. Invoke the function and save method on the PIL image. Use the download method from the files module to download the PIL image to a local drive.

Summary

The first experiment demonstrates an ANAS implementation of NST with an end-to-end code experiment. The second experiment demonstrates an ANAS implementation of NST with a pre-trained transfer model from the TensorFlow Lite module.

CHAPTER 13

Object Detection

Object detection is an automated computer vision technique for locating instances of objects in digital photographs or videos. Specifically, object detection draws bounding boxes around one or more effective targets located in a still image or video data. An **effective target** is the object of interest in the image or video data that is being investigated. The effective target (or targets) should be known at the beginning of the task.

Object Detection in a Natural Scene

Detecting objects in natural scenes is effortless for us, but extremely challenging for computer algorithms. As humans, we look at a scene and identify objects of interest without a thought. We process visual data in the ventral visual stream, which is a hierarchy of areas in the brain that helps in object recognition. We recognize different types and sizes of objects and categorize them with the aid of cells in our visual cortex that respond to simple shapes like lines and curves.

With object detection, data scientists attempt to mimic what humans do with computer algorithms. But before a computer algorithm can even begin to detect objects, a natural scene must be captured as a digital image (or digital video). A **digital image** is composed of picture elements (or pixels). A **pixel** is the basic logical unit in digital graphics. Simply, a pixel is a tiny square of color.

Pixels are uniformly combined in a two-dimensional (2D) grid to form a complete digital image, video, text or any visible element on a computer display. Each pixel has a specific number, and this number tells the algorithm its color. The number represents the pixel intensity value. The pixel intensity value represents a grayscale or color image.

In grayscale images, each pixel represents the intensity of only one color. So a grayscale image can be represented by a single 2D matrix (or grid). In the standard RGB system, color images have three channels (red, green, and blue). So a color image can be

D. Paper, *State-of-the-Art Deep Learning Models in TensorFlow*, https://doi.org/10.1007/978-1-4842-7341-8_13

represented by three 2D matrices. One matrix represents the intensity of red in the pixel. One represents the intensity of green in the pixel. And one represents the intensity of blue in the pixel.

Each matrix is considered a color channel. A **color channel** is an array of values (one per pixel) that together specify one aspect or dimension of the image. So RGB color contains three color channels - red, green, and blue. Each color channel is expressed from 0 (least saturated) to 255 (most saturated). So 16,777,216 (256^3) different colors can be represented in the RGB color space by combining red, green, and blue pixel intensities.

Once a digital image is captured as a grayscale (2D matrix) or color (three color channel 2D matrices) image, it is processed for model consumption. Only then can computer algorithms inspect each 2D grid of pixel values to begin identifying patterns.

Detection vs. Classification

Image classification presents an input image to a neural network so it can learn a single class label associated with the image. The network can also learn the probability associated with the class label. The learned class label characterizes the contents of the entire image or at least the most dominant and visible contents of the image. So an image classification network is able to learn how to correctly label an image of a cat.

Object detection presents an input image to a neural network so it can learn the exact location of an image in a picture object (or scene). To locate input images in a scene, object detection algorithms create a list of bounding boxes for each object in the picture. Bounding boxes are created as the (x, y) coordinate input image locations. The algorithms also identify the class labels associated with each bounding box and the probability/confidence score associated with each bounding box and class label.

Simply, image classification involves one image into the network and one class label out. Object detection involves one image into the network and multiple bounding boxes and their associated class labels out.

Object detection is commonly applied to count, locate, and label objects in a scene with precision. Neural networks are the state-of-the-art methods for object detection. Convolutional neural networks are frequently used to automatically learn inherent features of an object to identify them within an image by differentiating it from its background.

Whereas image classification involves assigning a class label to an image, object localization involves drawing a bounding box around one or more objects in an image scene. So classification separates images into different classes, while object localization identifies effective targets in an image scene. Object detection is more challenging than localization because it combines classification and object localization to learn how to draw a bounding box around each object of interest in the image (or effective target) and assign it a class label. The difference between object localization and object detection is subtle. Object localization aims to locate the main (or most visible) object in an image scene, while object detection tries to find out all the objects and their boundaries.

Imagine that an image contains two cats and a person. Object detection networks are able to locate instances of entities and classify the types of entities found within the image. So an object detection network can locate two cats and a person in this image and classify them correctly.

Object detection is commonly confused with image recognition. So how are they different? Image recognition assigns a label to an image. A picture of a dog receives the label *dog*. A picture of two dogs still receives the label *dog*. Object detection, however, draws a bounding box around each dog and labels the box as *dog*. The model predicts where each object is in the image and the label that should be applied to it. So object detection provides more information about an image than recognition.

Bounding Boxes

A **bounding box** is an imaginary rectangle that serves as a point of reference for object detection and creates a collision box for that object. The area of a bounding box (usually shortened to *bbox*) is defined by two longitudes and two latitudes where latitude is a decimal number between -90.0 and 90.0 and longitude is a decimal number between -180.0 and 180.0. Data annotators draw bbox rectangles to outline an object of interest within each image (scene) by defining its x and y coordinates.

A hitbox (or collision box) is an invisible shape commonly used in video games for real-time collision detection. So it is a type of bounding box. It is often a rectangle (in 2D games) or cuboid (in 3D) that is attached to and follows a point on a visible object (such as a model or a sprite).

Basic Structure

Deep learning object detection models typically have two parts. An *encoder* takes an image as input and runs it through a series of blocks and layers that learn to extract statistical features used to locate and label objects. Outputs from the encoder are then passed to a *decoder*, which predicts bounding boxes and labels for each object.

The simplest decoder is a pure regressor. A regressor is the name given to any variable in a regression model that is used to predict a response variable. A regressor is also referred to as an explanatory variable, an independent variable, a manipulated variable, a predictor variable or a feature. The whole point of building a regression model is to understand how changes in a regressor lead to changes in a response variable (or regressand). So a regressor is a feature, and a regressand is a response variable (or target).

Regression is a supervised ML technique used to predict continuous values. The goal of a regression algorithm is to plot a best-fit line or a curve between the data. A regressor is connected to the output of the encoder and directly predicts the location and size of each bounding box.

The output of a regressor model is the x, y coordinate pair for the object and its extent (or intensity) in the image. But a regressor is limited because we need to specify the number of boxes ahead of time. If an image has two dogs but the regressor model was designed to detect a single object, one of the dogs goes unlabeled. But if we know the number of objects we need to predict in each image ahead of time, pure regressor-based models may be a good option.

An extension of the regressor approach is a region proposal network. The decoder in a region proposal network proposes regions of an image where it believes an object might reside. The pixels belonging to these regions are then fed into a classification subnetwork to determine a label (or reject the proposal). The network then runs the pixels containing those regions through a classification network. The benefit of this method is a more accurate and flexible model that can propose arbitrary numbers of regions that may contain a bounding box. But the added accuracy comes at the cost of computational efficiency.

A single shot detector (SSD) seeks a middle ground. Rather than using a subnetwork to propose regions, a SSD relies on a set of predetermined regions. A grid of anchor points is laid over the input image. At each anchor point, boxes of multiple shapes and sizes serve as regions. For each box at each anchor point, the model outputs a prediction of whether or not an object exists within the region and modifications to the box's location and size to make it fit the object more precisely. Because there are multiple boxes at each anchor point and anchor points may be close together, a SSD produces many potential detections that overlap. So post-processing must be applied to SSD outputs to cull the best one by pruning deficient ones. The most popular post-processing technique for SDD is non-maximum suppression. *Non-max suppression* is used to select the most appropriate bounding box for an object.

Object detectors output the location and label for each object, but how do we measure performance? The most common metric for object location is intersection-over-union (IOU). Given two bounding boxes, IOU computes the area of the intersection and divides by the area of the union. The value ranges from 0 (no interaction) to 1 (perfectly overlapping). A simple percent correct metric can be used for labels.

Notebooks for chapters are located at the following URL:

https://github.com/paperd/deep-learning-models

We demonstrate object detection with an end-to-end code experiment. Begin setting up the Colab ecosystem by importing the main TensorFlow library and instantiating the GPU.

Import the TensorFlow Library

Import the library and alias it as **tf**:

```
import tensorflow as tf
```

Aliasing the TensorFlow library as tf is common practice.

GPU Hardware Accelerator

As a convenience, we provide the steps to enable the GPU in a Colab notebook:

1. Click *Runtime* in the top-left menu.
2. Click *Change runtime type* from the drop-down menu.
3. Choose *GPU* from the *Hardware accelerator* drop-down menu.
4. Click *Save*.

Verify that the GPU is active:

```
tf.__version__, tf.test.gpu_device_name()
```

If '/device:GPU:0' is displayed, the GPU is active. If '..' is displayed, the regular CPU is active.

Note If you get the error **NAME 'TF' IS NOT DEFINED**, re-execute the code to import the TensorFlow library!

Object Detection Experiment

We grab images from Google Drive and Wikimedia Commons for the experiment. We use the pre-trained *faster_rcnn/openimages_v4/inception_resnet_v2* object detection module for image object detection. The object detection model is trained on Open Images V4 (version 4) with the ImageNet pre-trained Inception ResNet V2 (version 2) as the image feature extractor. The module internally performs non-maximum suppression. The maximum number of detections outputted is 100. So it detects a maximum of 100 objects in a given scene. Detections are outputted for 600 boxable categories. A *boxable category* is one that is capable of (or suitable for) placing in a bounding box.

Open Images is a dataset of approximately nine million richly annotated images with image-level labels, object bounding boxes, object segmentations, visual relationships, and local narratives. The images are very diverse and often contain complex scenes with several objects (8.4 per scene on average).

The training set of *Open Images V4* contains 14.6 million bounding boxes for 600 object classes on 1.74 million images, which makes it the largest existing dataset with object location annotations (as of this writing). The boxes have been largely manually drawn by professional annotators to ensure accuracy and consistency. The images are very diverse and often contain complex scenes with several objects (8.4 per image on average). Moreover, the dataset is annotated with image-level labels spanning thousands of classes.

Note It is recommended to run this module on a GPU to get acceptable inference times.

Import Requisite Libraries

Enable access to the TF-hub module:

```
import tensorflow_hub as hub
```

Access a plotting module:

```
import matplotlib.pyplot as plt
```

Access modules for file handling and manipulating data in memory:

```
import tempfile
from six.moves.urllib.request import urlopen
from six import BytesIO
```

This *tempfile* module creates temporary files and directories. The *urlopen* module is the uniform resource locator (URL) handling module for Python. It is used to fetch a URL. A URL is a reference to a web resource that specifies its location on a computer network and a mechanism for retrieving it. The *BytesIO* module manipulates byte data in memory.

Access modules from the PIL library:

```
from PIL import Image
from PIL import ImageColor
from PIL import ImageDraw
from PIL import ImageFont
from PIL import ImageOps
```

The *Image* module allows image upload into memory. The remaining modules enable image manipulation.

Import NumPy:

```
import numpy as np
```

Create Functions for the Experiment

The first function displays an image:

```
def display_image(image):
  fig = plt.figure(figsize=(20, 15))
  plt.grid(False)
  plt.imshow(image)
  plt.axis('off')
```

We present many different display functions to provide alternative ways to display images. For practice, create your own display function (or functions) and use it for this experiment.

The second function draws a bounding box around an image within a scene as shown in Listing 13-1.

Listing 13-1. Function to Draw Bounding Boxes

```
def draw_bounding_box_on_image(
    image, ymin, xmin, ymax, xmax,
    color, font, thickness=4, display_str_list=()):
  """Adds a bounding box to an image."""
  draw = ImageDraw.Draw(image)
  im_width, im_height = image.size
  (left, right, top, bottom) = (
      xmin * im_width, xmax * im_width,
      ymin * im_height, ymax * im_height)
  draw.line([(left, top), (left, bottom),
             (right, bottom), (right, top),
             (left, top)],
             width=thickness, fill=color)
  display_str_heights = [font.getsize(ds)[1]
                         for ds in display_str_list]
  total_display_str_height = (
      1 + 2 * 0.05) * sum(display_str_heights)
  if top > total_display_str_height:
    text_bottom = top
```

```
  else:
    text_bottom = top + total_display_str_height
  for display_str in display_str_list[::-1]:
    text_width, text_height = font.getsize(display_str)
    margin = np.ceil(0.05 * text_height)
    draw.rectangle(
        [(left, text_bottom - text_height - 2 * margin),
         (left + text_width, text_bottom)], fill=color)
    draw.text(
        (left + margin, text_bottom - text_height - margin),
        display_str, fill='black', font=font)
    text_bottom -= text_height - 2 * margin
```

The function accepts a processed image, x and y minimum and maximum coordinates, color, font, thickness, and a list of display strings. The processed image is sent by the *draw_boxes* function, which is presented next. The x and y coordinates provide the boundaries of the bounding box for the image. The color, font, thickness, and list of strings are provided by the *draw_boxes* function. The remainder of the function adds a bounding box to an image.

Begin by creating an ImageDraw object and assigning it to variable *draw*. The *ImageDraw* module is used to create new images, annotate or retouch existing images, and generate graphics on the fly for web use. Image x and y coordinates are assigned to variables, and boundary lines are assigned to the ImageDraw object.

The function continues by assigning the list of display strings to variable *display_str_heights*. Display strings are used to label the bounding boxes for each image in a scene. If the total height of the display strings added to the top of the bounding box exceeds the top of the image, we need to stack the strings below the bounding box instead of above it. The reason is to ensure that strings that identify each image in a bounding box are readable. The *total_display_str_height* variable ensures that each display string has a reasonable top and bottom margin for display. You can experiment with this value, but the setting is pretty good as it is. The next bit of logic checks the top and bottom margins to ensure a good fit of the image inside the bounding box container.

The remainder of the function reverses (with a for loop) the list of display strings for display from bottom to top. The *draw.rectangle* method draws a bounding box around each image. The *draw.text* method labels each bounding box with its appropriate label. When the loop is finished, an image that includes all bounding boxes is created.

The third function accepts an image, boxes, class labels, scores, maximum number of boxes, and minimum score. The parameters with the exception of the image are generated by a pre-trained model. With the accepted parameter values, it overlays labeled boxes on an image with formatted scores and label names as shown in Listing 13-2.

Listing 13-2. Container Function for Drawing Bounding Boxes

```
def draw_boxes(
    image, boxes, class_names, scores,
    max_boxes=10, min_score=0.1):
  # Overlay labeled boxes on an image with formatted scores and label
    names.
  colors = list(ImageColor.colormap.values())
  one = '/usr/share/fonts/truetype/liberation/'
  two =  'LiberationSansNarrow-Regular.ttf'
  font_url = one + two
  try:
    font = ImageFont.truetype(font_url, 25)
  except IOError:
    print('Font not found, using default font.')
    font = ImageFont.load_default()
  for i in range(min(boxes.shape[0], max_boxes)):
    if scores[i] >= min_score:
      ymin, xmin, ymax, xmax = tuple(boxes[i])
      display_str = '{}: {}%'.format(
          class_names[i].decode('ascii'),
          int(100 * scores[i]))
      color = colors[hash(class_names[i]) % len(colors)]
      image_pil = Image.fromarray(
          np.uint8(image)).convert('RGB')
      draw_bounding_box_on_image(
          image_pil, ymin, xmin, ymax, xmax,
          color, font, display_str_list=[display_str])
      np.copyto(image, np.array(image_pil))
  return image
```

The draw_boxes function is a container because it accepts the image and dictionaries from the pre-trained model, calls *draw_bounding_box_on_image*, and returns an image with bounding boxes drawn around detected objects.

Although the function looks complex, it is pretty straightforward. It creates the colors inherent in the image. We need the colors because we recreate the detected objects from the scene so we can draw bounding boxes around them. It then creates the fonts that we use to label each bounding box. For practice, you can change the fonts. It then loops through all of the objects to retrieve (from dictionaries output from a pre-trained model) their x and y coordinates in the scene, labels, scores, image pixels, and colors to supply to the *draw_bounding_box_on_image* function. This function draws the bounding boxes for each object and returns a scene with bounding boxed images.

Note To avoid confusion, think of the original image as a scene. The pre-trained model learns how to identify image objects in the scene. It outputs a set of dictionaries that contain information about the objects it detects and coordinates for the bounding boxes from the scene. We then draw the bounding boxes from the dictionary information. So when we talk about image objects, we mean the objects that are detected from the scene.

Load a Pre-trained Object Detection Model

Load an object detection module and apply it on the downloaded image:

```
p1 = 'https://tfhub.dev/google/faster_rcnn/'
p2 = 'openimages_v4/inception_resnet_v2/1'
URL = p1 + p2
module_handle = URL
obj_detect = hub.load(module_handle).signatures['default']
```

The pre-trained model is an object detection model trained on Open Images V4 with ImageNet pre-trained Inception ResNet V2 as the image feature extractor.

Open Images is huge! It contains 15,851,536 boxes on 600 categories, 2,785,498 instance segmentations on 350 categories, 3,284,280 relationship annotations on 1,466 relationships, 675,155 localized narratives, and 59,919,574 image-level labels on 19,957 categories. It also includes an optional extension with 478,000 crowdsourced images with more than 6,000 categories.

The module performs non-maximum suppression internally. The maximum number of detections outputted is 100. Detections are outputted for 600 boxable categories. It is recommended to run this module on a GPU to get acceptable inference times.

The model accepts a variable-size three-channel image. It outputs several dictionaries including detection_boxes (bounding box coordinates), detection_class_entities (detection class names), detection_class_names (human-readable class names), detection_class_labels (labels as tensors), and detection_scores (detection scores). Detection scores indicate how confident the model is in labeling the object.

Note We provide the details on Open Images for the curious. The pre-trained model handles the workload. We just use its outputs to create a visualization.

Load an Image from Google Drive

Mount Google Drive to the Colab notebook:

```
from google.colab import drive
drive.mount('/content/gdrive')
```

Be sure that the image is in the *appropriate directory* (i.e., Colab Notebooks) in your Google Drive!

Access and display the image:

```
img_path = 'gdrive/My Drive/Colab Notebooks/images/cats_dogs.jpg'
pil_image = Image.open(img_path)
display_image(pil_image)
```

Convert the JPEG image to a Python Imaging Library (PIL) image and display it. PIL is a library that supports opening, manipulating, and saving many different image file formats. It is also known as the Pillow library.

Check image size:

```
pil_image.size
```

Prepare the Image

Generate a temporary path for the image file:

```
_, filename = tempfile.mkstemp(suffix='.jpg')
filename
```

Prepare the image for processing and save it to the temporary file path:

```
pil_image_rgb = pil_image.convert('RGB')
pil_image_rgb.save(filename, format='JPEG', quality=90)
print('Image downloaded to %s.' % filename)
display_image(pil_image)
```

Run Object Detection on the Image

Create a function to load the image:

```
def load_img(path):
  img = tf.io.read_file(path)
  img = tf.image.decode_jpeg(img, channels=3)
  return img
```

The function loads the image and prepares it for the pre-trained model.

Create a function to run object detection as shown in Listing 13-3.

Listing 13-3. Object Detection Function

```
def run_detector(detector, path):
  img = load_img(path)
  converted_img  = tf.image.convert_image_dtype(
      img, tf.float32)[tf.newaxis, ...]
  result = detector(converted_img)
  result = {key:value.numpy()
            for key,value in result.items()}
  print("Found %d objects." %\
        len(result["detection_scores"]))
  image_with_boxes = draw_boxes(
      img.numpy(), result["detection_boxes"],
```

```
    result["detection_class_entities"],
    result["detection_scores"])
  display_image(image_with_boxes)
```

The function accepts the loaded pre-trained model signature and path to the image (scene). It loads the image and converts it to a NumPy array for model consumption. Run the model signature on the NumPy image. Retrieve the dictionary key/value pairs from dictionaries outputted from the pre-trained model. Create a scene with bounding boxes around detected objects and display it.

Run the detector:

```
run_detector(obj_detect, filename)
```

As we know, the pre-trained model is limited to detecting 100 objects. But the message is misleading because it says that 100 objects are found no matter what scene is fed into the model.

Detection is perfect, but don't get too excited. The model is powerful enough to detect images in a simple scene. The scene is simple because the background offers no distractions and each dog/cat is separately presented.

Let's try another one:

```
img_path = 'gdrive/My Drive/Colab Notebooks/images/butterfly.jpg'
pil_image = Image.open(img_path)
display_image(pil_image)
```

Process the image:

```
_, filename = tempfile.mkstemp(suffix='.jpg')
pil_image_rgb = pil_image.convert('RGB')
pil_image_rgb.save(filename, format='JPEG', quality=90)
print('Image downloaded to %s.' % filename)
```

Run the detector:

```
run_detector(obj_detect, filename)
```

Detection is perfect, but the image is simple.

Detect Images from Complex Scenes

Let's try detection on more complex images. We've already located some images from Wikimedia Commons, but you can locate your own by following a few simple steps:

1. Go to the following URL: *https://commons.wikimedia.org/wiki/Main_Page*
2. Click the Images link.
3. Click an image.
4. Right-click the image.
5. Select "Copy link address" from the drop-down menu.
6. Paste the link address into a code cell.
7. Surround the link address with single or double quotes.
8. Assign to a variable.

Create a Download Function

Create a function to download, process, and save an image to a temporary file path as shown in Listing 13-4.

Listing 13-4. Download and Preprocess Function

```
def download_and_resize_image(
    url, new_width=256, new_height=256,
    display=False):
  _, filename = tempfile.mkstemp(suffix='.jpg')
  response = urlopen(url)
  image_data = response.read()
  image_data = BytesIO(image_data)
  pil_image = Image.open(image_data)
  pil_image = ImageOps.fit(
      pil_image, (new_width, new_height),
      Image.ANTIALIAS)
  pil_image_rgb = pil_image.convert('RGB')
  pil_image_rgb.save(
```

```
      filename, format='JPEG', quality=90)
  print('Image downloaded to %s.' % filename)
  if display:
    display_image(pil_image)
  return filename
```

The function generates a temporary path for the image file. It then reads the image file from the supplied URL. The function continues by converting the image file to a PIL image. The PIL image is then resized, converted to RGB, and saved to the temporary file path. The function ends by returning the filename of the PIL image.

Load an Image Scene

Load a scene from a Wikimedia Commons URL:

```
p1 = 'https://upload.wikimedia.org/wikipedia/commons/7/79/'
p2 = 'At_taverna_under_the_church%2C_Ano_Potamia%2C_Naxos%'
p3 = '2C_190574.jpg'
URL = p1 + p2 + p3

downloaded_image_path = download_and_resize_image(
    URL, 1280, 856, True)
```

The source for the image scene is located at *https://commons.wikimedia.org/wiki/File:At_taverna_under_the_church,_Ano_Potamia,_Naxos,_190574.jpg*

Detect

Run object detection:

```
run_detector(obj_detect, downloaded_image_path)
```

With a more complex scene, detection is not perfect. But it does make some correct detections.

Detect on More Scenes

Piece together some paths as shown in Listing 13-5.

Listing 13-5. Image Paths

```
p1 = 'https://upload.wikimedia.org/wikipedia/commons/4/45/'
p2 = 'Green_Dragon_Tavern_%2836196%29.jpg'
tavern = p1 + p2

p1 = 'https://upload.wikimedia.org/wikipedia/commons/3/31/'
p2 = 'Circus_Circus_Hotel-Casino_sign.jpg'
casino = p1 + p2

p1 = 'https://upload.wikimedia.org/wikipedia/commons/9/91/'
p2 = 'Leon_hot_air_balloon_festival_2010.jpg'
balloon = p1 + p2

p1 = 'https://upload.wikimedia.org/wikipedia/commons/d/d8/'
p2 = '2012_Festival_of_Sail_-_7943922284.jpg'
sail = p1 + p2

p1 = 'https://upload.wikimedia.org/wikipedia/commons/a/ab/'
p2 = '17_mai_2018.jpg'
flag = p1 + p2

p1 = 'https://upload.wikimedia.org/wikipedia/commons/4/43/'
p2 = 'Fruit_baskets.jpg'
basket= p1 + p2

p1 = 'https://upload.wikimedia.org/wikipedia/commons/c/c7/'
p2 = 'Fruit_stands%2C_Rue_de_Seine%2C_Paris_22_May_2014.jpg'
stand= p1 + p2

p1 = 'https://upload.wikimedia.org/wikipedia/commons/9/95/'
p2 = 'Wine_tasting_%40_brown_brothers.jpg'
wine = p1 + p2
```

Create a function to detect images in a scene:

```
def detect_img(image_url):
  image_path = download_and_resize_image(image_url, 640, 480)
  run_detector(obj_detect, image_path)
```

Run object detection on one of the scenes:

```
detect_img(wine)
```

So scene complexity limits detection accuracy.

Try one more:

```
detect_img(sail)
```

Find the Source

We sometimes come across Wikimedia Commons images in our research. But the sources of such images are never (at least in our experience) included. If we want to use the image in any way, we must locate its source to see if it is permitted.

Find the Source of a Wikipedia Commons Image

We can find the source of an image with a few steps:

1. Substitute *commons* for *upload.*
2. Change *wikipedia* to *wiki.*
3. Substitute *commons/(number)/(number)* for *File:*
4. Translate the *%(number)* to its *HTML encoded equivalent.*

To find the HTML encoded equivalent, peruse *https://krypted.com/utilities/html-encoding-reference/*

Note We cannot guarantee that this process works for every image, but it works for the ones that we have used.

Let's try the process on the tavern image:

https://upload.wikimedia.org/wikipedia/commons/4/45/Green_Dragon_Tavern_%2836196%29.jpg

Substitute *commons* for *upload*:

https://commons.wikimedia.org/wikipedia/commons/4/45/Green_Dragon_Tavern_%2836196%29.jpg

Change *wikipedia* to *wiki*:

https://commons.wikimedia.org/wiki/commons/4/45/Green_Dragon_Tavern_%2836196%29.jpg

Substitute *commons/(number)/(number)* for *File*:

https://commons.wikimedia.org/wiki/File:Green_Dragon_Tavern_%2836196%29.jpg

Translate:

https://commons.wikimedia.org/wiki/File:Green_Dragon_Tavern_(36196).jpg

The %28 and %29 codes translate into left and right parentheses from the HTML Encoding Reference.

The casino image is easier because we don't have to do any translations:

https://commons.wikimedia.org/wiki/File:Circus_Circus_Hotel-Casino_sign.jpg

Here are the sources for the remaining images:

https://commons.wikimedia.org/wiki/File:Leon_hot_air_balloon_festival_2010.jpg

https://commons.wikimedia.org/wiki/File:2012_Festival_of_Sail_-_7943922284.jpg

https://commons.wikimedia.org/wiki/File:17_mai_2018.jpg

https://commons.wikimedia.org/wiki/File:Fruit_baskets.jpg

https://commons.wikimedia.org/wiki/File:Fruit_stands,_Rue_de_Seine,_Paris_22_May_2014.jpg

https://commons.wikimedia.org/wiki/File:Wine_tasting_@_brown_brothers.jpg

Summary

We used a powerful pre-trained image detector model on several image scenes to demonstrate object detection. The model worked extremely well on simple scenes, but not as well on complex ones. As object detection in deep learning continues to evolve, we are confident that future models will vastly improve detection capability.

CHAPTER 14

An Introduction to Reinforcement Learning

Reinforcement learning (RL) is an area of machine learning that focuses on teaching intelligent agents how to take actions in an environment in order to maximize cumulative reward. **Cumulative reward** in RL is the sum of all rewards as a function of the number of training steps.

We train machine learning (ML) models by using rewards and punishments. When an agent makes a correct decision, we reward it with a positive point. With a wrong decision, we punish it with a negative point. From these responses, the model learns how to react in that particular situation (or environment). So the idea behind RL is that an agent learns from an environment by interacting with it and receiving rewards (and punishments) for performing actions.

Learning from interaction with the environment comes from our natural experiences. Imagine you're a child in a living room. You see a fireplace and you approach it. It's warm, it's positive, and you feel good. You understand that fire is a positive thing. But then you touch the fire. Ouch! It burns your hand. From your interactions, you learn that fire is positive when you are a sufficient distance away because it produces warmth. But get too close and it burns. So humans learn through interaction with their environment.

RL is just a computational approach of learning from action. If this scenario was a RL experiment, the agent would get a positive reward of +1 for getting warm by the fire, but a negative reward of -1 for getting burnt by it.

RL is one of three basic ML paradigms alongside supervised learning and unsupervised learning. Supervised learning models learn from a labeled dataset with guidance. Unsupervised learning models learn from unlabeled data without guidance. RL models learn by trial and error when an agent interacts with an environment, performs actions, and is rewarded if the action is correct or punished if it is incorrect.

D. Paper, *State-of-the-Art Deep Learning Models in TensorFlow*, https://doi.org/10.1007/978-1-4842-7341-8_14

The three components of reinforcement learning are agent, environment, and actions. An *environment* is a problem to be solved. An *agent* is an algorithm that interacts with the environment to solve a problem. *Actions* are an agent's interactions with the environment.

Laconically, the agent receives observations from an environment and takes actions. The environment uses rewards or punishments as signals for positive or negative behavior based on the actions taken by the agent. So the agent learns how to solve a problem by trial and error interactions with the environment.

Although the designer sets the reward policy, they give the learning model no hints or suggestions on how to solve the problem. The model learns *on its own* how to maximize reward through trial and error interactions between the agent and environment.

A **reward policy** is the set of rules that maximizes the reward function for a given reinforcement learning environment. So a *reward function* stipulates what the designer wants the agent to accomplish.

Challenges of Reinforcement Learning

The main challenge of RL is *creating the environment.* Creating an effective environment has three issues.

First, the environment is highly dependent on the problem to be solved. So it is contextually specific. The context (of the problem domain) drives design of each reinforcement learning environment to solve a specific set of tasks. Simply, a new environment must be designed for each new RL task. For instance, an environment for a self-driving car is not transferable to one for a self-flying drone. In contrast, a supervised learning classification model can be reused by other classification tasks.

Second, the reward policy is finite and structured. Environments for chess, Go, and Atari games are relatively simple. No matter how complex these games might appear to be, their rules are structured and finite. Even an extremely complex environment for a self-driving car is finite and structured. Although such an environment must deal with many unknowns, its reward policy is created by designers. Even the most brilliant designers on the planet can't create an infinite reward policy. For example, a self-driving car may be designed to work well in a variety of conditions, but it may not work in a different situation than that in which it was designed. A self-driving car may work when traveling a specific route, but what if construction calls for a detour but the detour signage has yet to be posted?

Third, the room for error where safety is a concern is pretty much zero. RL experiments in the healthcare industry come to mind. Such models must be tested and tweaked in many stages before they are ready for even simple testing. Transferring the model out of the testing phase into to the real world is where the real work begins.

Scaling and tweaking the neural network (or other ML models) controlling the agent is another challenge. The only way to communicate with the network is through its system of rewards and penalties. With this single conduit comes the possibility of catastrophic forgetting (or catastrophic interference). **Catastrophic forgetting** is the tendency of a neural network to completely and abruptly forget previously learned information upon learning new information. So when a neural network acquires new information, some of the old information is erased from the network. Catastrophic forgetting occurs because many of the weights (where information is stored) of the neural network are changed when new information is learned, which makes it less likely that prior knowledge is kept intact. During sequential learning, new inputs coming into a neural network can erase original input weights.

Another challenge is reaching a local optimum. A **local optimum** is the best solution to a problem within a small neighborhood of possible solutions. In contrast, a **global optimum** is the optimal solution when *every* possible solution is considered. A local optimum would be walking when the goal is to learn how to move (walk, run, and jump). In this case, the agent performs the task in a suboptimal (locally optimal) way, but not in an (globally) optimal way.

A final challenge is securing talented designers. A competent data scientist may not be hard to find, but the average salary can be well over $150,000 per year. Moreover, it may be difficult to find one with experience in creating a specific type of environment because environments are contextually specific.

Notebooks for chapters are located at the following URL:

https://github.com/paperd/deep-learning-models

We demonstrate reinforcement learning with a code experiment using a *very simple* RL environment. Begin setting up the Colab ecosystem by importing the main TensorFlow library and instantiating the GPU.

Import the TensorFlow Library

Import the library and alias it as **tf**:

```
import tensorflow as tf
```

Aliasing the TensorFlow library as tf is common practice.

GPU Hardware Accelerator

As a convenience, we provide the steps to enable the GPU in a Colab notebook:

1. Click *Runtime* in the top-left menu.
2. Click *Change runtime type* from the drop-down menu.
3. Choose *GPU* from the *Hardware accelerator* drop-down menu.
4. Click *Save*.

Verify that the GPU is active:

```
tf.__version__, tf.test.gpu_device_name()
```

If '/device:GPU:0' is displayed, the GPU is active. If '..' is displayed, the regular CPU is active.

Note If you get the error **NAME 'TF' IS NOT DEFINED**, re-execute the code to import the TensorFlow library!

Reinforcement Learning Experiment

We demonstrate a simple RL experiment with Cart-Pole. *Cart-Pole* is a game in which a pole is attached by an unactuated joint to a cart that moves along a frictionless track. The goal of the game is to keep the pole vertically upright. The starting state is randomly initialized between -0.05 and 0.05. The starting state includes the cart position, cart velocity, pole angle, and pole velocity. The pole velocity is measured at the tip of the pole. The Cart-Pole game is in 2D space. So the cart can only move left and right to balance the pole.

The goal of the experiment is to teach an agent how to balance a pole on a cart with a RL model. We can create our own environment to train the agent, but we don't have to do this because environments to tackle this simple pole balancing problem already exist. We use an environment from OpenAI Gym. *OpenAI Gym* is a toolkit that provides a wide variety of simulated environments including Atari games, board games, and (2D and 3D) physical simulations.

Install and Configure OpenAI Gym on Colab

Most of the requirements of Python packages (for OpenAI Gym) are already fulfilled on Colab. But we still need to install dependencies:

```
!pip install gym
!apt-get install python-opengl -y
!apt install xvfb -y
```

We need gym, opengl, and xvfb dependencies to enable the Python libraries for OpenAI Gym, graphics libraries for visualizations, and a display server for rendering graphics. *Gym* is an open source interface for RL tasks. It supports teaching agents everything from walking to playing games like Pong or Pinball.

OpenGL is a graphics library supported by multiple platforms including Windows, Linux, and MacOS. *Xvfb* (X virtual framebuffer) is a display server implementing the X11 display server protocol. It runs in memory and does not require a physical display.

We also need to install dependencies for displaying output rendered from the environment:

```
!pip install pyvirtualdisplay
!pip install piglet
```

The *pyvirtualdisplay* library enables virtual display. The *piglet* library provides an object-oriented API for the creation of games and other multimedia applications.

Import Libraries

Import and activate the virtual display library:

```
import pyvirtualdisplay

display = pyvirtualdisplay.Display(
    visible=0, size=(1400, 900)).start()
```

We used 1400 × 900 for our virtual display. The size we chose is arbitrary. Feel free to experiment with the size of the display.

Import the *gym* library:

```
import gym
```

The gym library is a collection of environments designed for testing and developing RL algorithms. It saves us from having to create complicated environments on our own.

Create an Environment

Create the *Cart-Pole* environment:

```
env = gym.make('CartPole-v1')
```

The Cart-Pole environment is a 2D simulation that accelerates a cart left or right to balance a pole placed on top of it. A pole is attached by an unactuated joint to a cart that moves along a frictionless track. The system is controlled by applying a force of +1 or -1 to the cart. The pendulum (or pole) starts upright, and the goal is to prevent it from falling over.

Initialize the environment:

```
env.seed(0)
obs = env.reset()
obs
```

Initialize the environment with the *reset* method. After it initializes, the method returns an observation.

Observations vary depending on the environment. In this case, an observation is a 1D NumPy array composed of four floats that represent the cart's horizontal position, velocity, angle of the pole (0 = vertical), and angular velocity. Any positive number indicates movement to the *right* for angle of the pole and angular velocity. Any negative number indicates movement to the *left*. For horizontal position, a negative number means that the pole is *tilting left* and a positive number *tilting right*. For velocity, a positive number means the cart is *speeding up* and a negative number *slowing down*.

The initial rendering state (with seed of 0) produces the 1D NumPy array:

array([-0.04456399, 0.04653909, 0.01326909, -0.02099827])

So the initial state of the pole is not completely horizontal (obs[0] is slightly negative), its velocity is slowly increasing (obs[1] is slightly positive), the pole is angled slightly to the right (obs[2] is slightly positive), and the angular velocity is going toward the left (obs[3] is slightly negative).

An environment can be visualized by calling its *render* method, and we can pick the rendering mode (the rendering options depend on the environment). Display the rendering environment at its initial state:

```
env.render()
```

In our case, the env.render() command displays *True*, which is the initial state because we just initialized it as such.

In RL, we must render (initialize) the environment to its beginning state as a precursor to training. The reason is that we want to begin the environment in its initial state to give no clues to the agent. We want the agent to learn how to solve the task from scratch.

Display the Rendering from the Environment

Set mode='rgb_array' to get an image of the environment as a NumPy array:

```
img = env.render(mode='rgb_array')
img.shape
```

The image shape is rendered from the Cart-Pole environment.

Create a function to display the rendered image of the pole position on the cart from the environment as configured in Listing 14-1.

Listing 14-1. Function to Display the Rendered Environment

```
def plot_environment(env, figsize=(5,4)):
  plt.figure(figsize=figsize)
  img = env.render(mode='rgb_array')
  plt.imshow(img)
  plt.axis('off')
  return img
```

Display:

```
import matplotlib.pyplot as plt

plot_environment(env)
plt.show()
```

We see the rendering from the environment in its initialized state. So the pole is very nearly vertical on the cart, but tilted slightly to the right.

Display Actions

Let's see how to interact with the environment we created. The agent needs to select an action from an action space. An **action space** is the set of possible actions that an agent can take.

Ask the environment about possible actions:

```
env.action_space
```

Discrete(2) means that the possible actions for the Cart-Pole environment are integers 0 and 1. Accelerating left is 0 and accelerating right is 1. So the environment's action space has two possible actions, which means that the agent can accelerate toward the left or toward the right. Of course, other environments may have additional discrete actions or other kinds of actions like continuous ones.

Note The Cart-Pole environment is the simplest one that can be created! Real-world reinforcement learning environments have enormous action spaces containing many possible actions.

Reset the environment and see how the pole is leaning by looking at its angle:

```
env.seed(0)
obs = env.reset()
indx = 2
obs[indx]
```

The third position (index of 2) in the *obs* array is the angle of the pole. If the value is below 0, the pole angles to the left. If above 0, it angles to the right. The value is barely over zero. So the pole is tilted slightly toward the right because obs[2] is > 0.

Note We didn't have to reset the environment again because we did so in the previous section. But an environment should be reset at the beginning of an experiment to ensure that the agent learns how to solve the task from scratch (with no clues). So we did this again just to instill a good RL habit.

As we already know, the Cart-Pole environment only has two actions, left (0) and right (1). Let's accelerate the cart toward the right by setting *action=1*:

```
action = 1
obs, reward, done, info = env.step(action)
print ('obs array:', obs)
print ('reward:', reward)
print ('done:', done)
print ('info:', info)
```

The *step* method executes the given action and returns four values. *obs* is the new observation. The cart is now moving toward the right because obs[1] > 0. The pole is still tilted toward the right because obs[2] > 0, but its angular velocity is now negative because obs[3] < 0. So it will likely be tilted toward the left after the next step.

In this simplest of environments, *reward* is always 1.0 at every step. So the goal is to keep the episode running as long as possible. The *done* value is True when the episode is over. The episode is over if the pole tilts too much or goes off the screen or we win the game. The *info* value provides extra information. In this case, there is no extra information. Once we finish using an environment, call the *close* method to free resources.

Note It is not necessary to free resources because our experiment is so simple. But it is a good idea to close the environment upon task completion for more complex RL tasks as they tend to consume a lot of computer resources.

The environment tells the agent each new observation, the reward, when the game is over, and information it got during the last step. Display the pole position:

```
plot_environment(env)
plt.show()
```

The pole is still tilted toward the right.

Display the reward that the agent received in the last step:

```
reward
```

Of course, the reward is 1.0 because it is always 1.0 in this simplest of experiments.

Test if the game is over:

```
done
```

Since the value is *False*, the game is not over. If the output is *True*, the task is completed so the game is over.

An **episode** is the sequence of steps from the moment the environment is reset until it is done. At the end of an episode (i.e., when the step method returns done=True), reset the environment before continuing to use it.

To automatically reset when an episode is over:

```
if done:
  obs = env.reset()
else:
  print ('game is not over!')
```

Note We just show you the preceding code for informational purposes. We show you when to use env.reset() in the context of the RL experiment.

Simple Neural Network Reward Policy

How can we make the pole remain upright? We need to define a reward policy. In action, a reward policy is simply the strategy the agent uses to select an action at each step. The agent can use all past actions and observations to decide what to do.

Let's create a neural network that takes observations as inputs and outputs the action to take for each observation. To choose an action, our network estimates a probability for each action and randomly selects an action based on the estimated probabilities. In the case of the Cart-Pole environment, there are just two possible actions (left and right). So we only need one output neuron that outputs the probability p of action 0 (left) and the probability $1 - p$ of action 1 (right).

Clear previous models and generate a seed:

```
import numpy as np

tf.keras.backend.clear_session()
tf.random.set_seed(0)
np.random.seed(0)
```

Determine the observation space:

```
obs_space = env.observation_space.shape
obs_space
```

The *observation space* is another term for the reward policy. The observation space (as shown earlier in the observation array) is a 1D NumPy array composed of four floats that represent the cart's horizontal position, velocity, angle of the pole (0 = vertical), and angular velocity. So the observation space is contained in a 1D NumPy array with 4 elements.

Note The terms policy, reward policy, observation array, and observation space are interchangeable.

Set the number of inputs for the policy network:

```
n_inputs = env.observation_space.shape[0]
n_inputs
```

Create the policy network:

```
from tensorflow.keras.models import Sequential
from tensorflow.keras.layers import Dense

model = Sequential([
  Dense(5, activation='elu', input_shape=[n_inputs]),
  Dense(1, activation='sigmoid')
])
```

The policy network is a simple *Sequential* neural network. The number of inputs is the size of the observation space, which is 4 in our case. We include only five neurons in the first layer because this is such a simple problem. We only need to output a single probability (the probability of going left). So we use sigmoid activation in the output layer to generate a single output neuron as a logit. We only need to output a single probability p because we can get the probability of going right with 1 – p. If we had more than two possible actions, we would still use one output neuron per action and substitute softmax activation in the output layer.

In this particular environment, past actions and observations can safely be ignored because each observation contains the environment's full state. If there were some hidden state, we might need to consider past actions and observations to try to infer the hidden state of the environment. For example, if the environment only revealed the position of the cart but not its velocity, we would have to consider not only the current observation but also the previous observation in order to estimate the current velocity. Another example is if the observations are noisy, we might want to use the past few observations to estimate the most likely current state. Our problem is as simple as can be because the current observation is noise-free and contains the environment's full state.

Why do we pick a random action based on the probability given by the policy network rather than just picking the action with the highest probability? Because this approach lets the agent find the right balance between exploring new actions and exploiting the actions that are known to work well.

Model Predictions

Create a function that runs the model to play one episode and returns the frames so we can display an animation as shown in Listing 14-2.

Listing 14-2. Function to Return the Frames from One Episode

```
def render_policy_net(model, n_max_steps=200, seed=0):
  frames = []
  env = gym.make('CartPole-v1')
  env.seed(seed)
  np.random.seed(seed)
  obs = env.reset()
  for step in range(n_max_steps):
    frames.append(env.render(mode='rgb_array'))
    left_proba = model.predict(obs.reshape(1, -1))
    action = int(np.random.rand() > left_proba)
    obs, reward, done, info = env.step(action)
    if done:
      break
  env.close()
  return frames
```

Establish the Cart-Pole environment and reset it. Create a loop to run a number of steps until the episode is over. Begin each step by appending the visualization of the environment rendering to the *frames* list. Continue by making an action prediction from the model. Next, establish an action based on the prediction. Execute the *step* method based on the action. Continue looping until the episode is over. End by returning the list of frames.

Create functions to show animation of the frames as shown in Listing 14-3.

Listing 14-3. Functions to Animate Frames

```
import matplotlib.animation as animation
import matplotlib as mpl

def update_scene(num, frames, patch):
  patch.set_data(frames[num])
  return patch,

def plot_animation(frames, repeat=False, interval=40):
  fig = plt.figure()
  patch = plt.imshow(frames[0])
```

```
    plt.axis('off')
    anim = animation.FuncAnimation(
        fig, update_scene, fargs=(frames, patch), blit=True,
        frames=len(frames), repeat=repeat, interval=interval)
    plt.close()
    return anim
```

The *plot_animation* function creates an animation by repeatedly calling the *update_scene* function. The plot_animation function accepts the frames list. The function continues by extracting an image (patch) from the frames list. It then calls update_scene to set the x and y coordinates for the patch. The animation is returned batched on the patch coordinates for each frame image.

Create the frames list based on the policy network:

```
frames = render_policy_net(model)
```

Animate

Create the animation:

```
anim = plot_animation(frames, interval=100)
```

Experiment with the *interval* parameter to see its impact on the animation. We set interval size to 100 just because we liked the result.

Note Increasing the interval just lengthens the number of seconds that the animation runs.

Render and display the animation. We show two ways to accomplish this. The first method uses the HTML library to display HTML elements:

```
from IPython.display import HTML
```

```
method1 = HTML(anim.to_html5_video())
method1
```

The animation is rendered to *html5 video* with the to_html5_video method and displayed with the HTML module.

The second method uses the runtime configuration library:

```
from matplotlib import rc

method2 = rc('animation', html='html5')
```

To implement the second method, just run the animation object:

```
anim
```

Ugh! The pole is falling to the left! The reason is that we've yet to implement a reward policy.

Implement a Basic Reward Policy

In the previous section, we just had the policy network run random predictions with *no interaction with the agent.* If the pole is tilting left, the agent has no way of knowing whether this is a bad or good action. We need to develop an environment with a reward policy that the agent can interact with to learn how to balance the pole on the cart.

Create a diverse environment space as shown in Listing 14-4.

Listing 14-4. Environment Space for the Experiment

```
n_environments = 50
n_iterations = 5000

envs = [gym.make(
    'CartPole-v1') for _ in range(n_environments)]
for index, env in enumerate(envs):
  env.seed(index)
np.random.seed(0)
observations = [env.reset() for env in envs]
optimizer = tf.keras.optimizers.RMSprop()
loss_fn = tf.keras.losses.binary_crossentropy
```

```
for iteration in range(n_iterations):
  target_probas = np.array(
      [([1.] if obs[2] < 0 else [0.])
      for obs in observations])
  with tf.GradientTape() as tape:
    left_probas = model(np.array(observations))
    loss = tf.reduce_mean(
        loss_fn(target_probas, left_probas))
    print('\rIteration: {}, Loss: {:.3f}'.\
          format(iteration, loss.numpy()), end='')
    grads = tape.gradient(loss, model.trainable_variables)
    optimizer.apply_gradients(
        zip(grads, model.trainable_variables))
    actions = (np.random.rand(n_environments, 1) >\
               left_probas.numpy()).astype(np.int32)
  for env_index, env in enumerate(envs):
    obs, reward, done, info = env.step(
        actions[env_index][0])
    observations[env_index] = obs if not done else env.reset()

for env in envs:
  env.close()
```

We create an environment space with 50 different environments in parallel to provide a diverse training batch at each step in the network. We train the network for 5,000 iterations. Of course, you can tweak the number of environments and iterations, but be cognizant of available computer resources.

We use the RMSprop optimizer because it seems to work pretty well. You can experiment with different optimizers to see how well the agent learns the task. We use binary cross-entropy for the loss function because we only have two discrete possible actions (left and right). The first line in the iteration loop checks the angle of the pole. If the angle < 0, the target action is left (proba(left) = 1.). Otherwise, the target action is right (proba(left) = 0).

Optimizers are algorithms (or methods) used to change the attributes of a neural network (such as weights and learning rate) to reduce loss. So optimizers solve optimization problems by minimizing the loss function.

For a complete list of TensorFlow optimizers, peruse

www.tensorflow.org/api_docs/python/tf/keras/optimizers

The gradient tape section records the agent's actions during training. The tilt of the pole is determined. The agent takes actions to steady the pole on the cart. The actions are then fed to the 50 environments that determine the rewards given to the agent. This process repeats 5,000 times.

Finally, the environments are reset to free resources. We train with a custom training loop so we can easily use the predictions at each training step to advance the environments.

Create the frames for an animation:

```
frames = render_policy_net(model)
```

We now have the frames based on what the agent learned during training. We use the frames to create an animation.

Animate:

```
anim = plot_animation(frames, repeat=True, interval=100)
anim
```

Voilà! The animation verifies that our RL model worked. That is, we taught the agent how to balance the pole on the cart. But can we do even better?

Reinforce Policy Gradient Algorithm

We have yet to demonstrate the real breakthrough of RL. In the previous section, the agent learned from a reward policy. But can the agent learn a better policy on its own?

We can use a reinforce policy gradient algorithm to automate agent learning. *Policy gradients* optimize the parameters of a policy by following the gradients toward higher rewards.

How Do Policy Gradients Work?

To use policy gradients, let the neural network policy play the game several times. At each step, compute the gradients that make the chosen action even more likely. But don't apply the gradients at this point.

After running several episodes, compute each action's advantage with a discount factor at each step. A *discount factor* is computed by evaluating an action based on the sum of all rewards that come after the action. If an action's advantage is positive, the action was probably good. So **now** apply the gradients to make the action more likely to be chosen in the future. If it is negative, apply the opposite gradients to make the action less likely to be chosen. Finally, compute the mean of all resultant gradient vectors and use it to perform a *gradient descent* step. The mean of all resultant gradient vectors is calculated by taking the mean of each gradient (or opposite gradient) multiplied by its action advantage. The end result is the creation of an effective policy gradient descent algorithm that minimizes the loss function.

Train the Model with Policy Gradients

Begin by creating functions to play a step, play multiple episodes, discount rewards, and normalize discounted rewards.

Create a function that plays one step as shown in Listing 14-5.

Listing 14-5. Function to Play a Single Step

```
def play_one_step(env, obs, model, loss_fn):
  with tf.GradientTape() as tape:
    left_proba = model(obs[np.newaxis])
    action = (tf.random.uniform([1, 1]) > left_proba)
    y_target = tf.constant(
        [[1.]]) - tf.cast(action, tf.float32)
    loss = tf.reduce_mean(loss_fn(
        y_target, left_proba))
  grads = tape.gradient(loss, model.trainable_variables)
  obs, reward, done, info = env.step(
      int(action[0, 0].numpy()))
  return obs, reward, done, grads
```

In the *GradientTape* block, call the model with a single observation. Reshape the observation so that it becomes a batch containing a single instance (the model expects a batch). Get a probability of going left by sampling a random float between 0 and 1 and checking if it is greater than the probability. The *action* is False if the probability is left_proba and True if the probability is 1 – left_proba. Cast this Boolean (True or False) to a number of 0 (left) or 1 (right) with the appropriate probabilities. We then define the target probabilities of going left (1 - the action) or right (the action). If the action is 0 (left), the target probability of going left is 1. If the action is 1 (right), the target probability is 0.

Continue by computing the loss and use the tape to compute the gradient of the loss with regard to the model's trainable variables. Tweak the gradients later depending on how good or bad that action turned out to be. Finally, play the selected action and return the new observation, the reward, whether the episode is over or not, and the gradients.

Create a function to play multiple episodes and return the rewards and gradients for each episode and each step as shown in Listing 14-6.

Listing 14-6. Play Multiple Episodes Function

```
def play_multiple_episodes(
    env, n_episodes, n_max_steps, model, loss_fn):
  all_rewards = []
  all_grads = []
  for episode in range(n_episodes):
    current_rewards = []
    current_grads = []
    obs = env.reset()
    for step in range(n_max_steps):
      obs, reward, done, grads = play_one_step(
          env, obs, model, loss_fn)
      current_rewards.append(reward)
      current_grads.append(grads)
      if done:
        break
    all_rewards.append(current_rewards)
    all_grads.append(current_grads)
  return all_rewards, all_grads
```

The function returns a list of reward lists by calling the *play_one_step* function for the number of desired steps. This list contains one reward list per episode. Each reward list contains one reward per step. The function also returns a list of gradient lists. This list contains one gradient list per episode. Each gradient list contains one tuple of gradients per step. Each tuple contains one gradient tensor per trainable variable.

Simply, the policy gradient algorithm uses the *play_multiple_episodes* function to play the game several times. It then goes back and looks at all the rewards to discount and normalize them.

Discount and Normalize the Rewards

To discount and normalize rewards, we discount the rewards and normalize them. The *discount factor* determines how important each reward is in the current state of the RL algorithm during training. The idea is to teach the agent how to rate the importance of each reward as it occurs during learning so it can be compared to future rewards in the training process. *Normalizing discounted rewards* makes the gradient steeper for rewards (positive reinforcement) and shallower for punishments (negative reinforcement). Simply, a reward triggers a steeper gradient, and a punishment triggers a shallower gradient when discounted rewards are normalized. A steeper gradient speeds minimization of the loss function, which is the goal of any ML algorithm.

The *discount_rewards* function discounts rewards as shown in Listing 14-7.

Listing 14-7. Discount Rewards Function

```
def discount_rewards(rewards, discount_rate):
  discounted = np.array(rewards)
  for step in range(len(rewards) - 2, -1, -1):
    discounted[step] += discounted[step + 1] * discount_rate
  return discounted
```

Verify that the function works:

```
discount_rewards([10, 0, -50], discount_rate=0.8)
```

We give the function three actions. After each action, there is a reward. The first reward is 10, the second is 0, and the third is -50. We use a discount factor of 80%. So the third action gets -50 (full credit for the last reward). But the second action only gets -40 (80% credit for the last reward). Finally, the first action gets 80% of -40 (-32) plus full credit for the first reward (+10) leading to a discounted reward of -22.

Note We gave the *discount_rewards* function three actions with a list (or vector) of three values, namely, [10, 0, -50]. We can change the number of actions by adding or removing values to or from the list.

The *discount_and_normalize_rewards* function normalizes discounted rewards as shown in Listing 14-8.

Listing 14-8. Normalize Discounted Rewards Function

```
def discount_and_normalize_rewards(
    all_rewards, discount_rate):
  all_discounted_rewards =\
    [discount_rewards(rewards, discount_rate)
    for rewards in all_rewards]
  flat_rewards = np.concatenate(all_discounted_rewards)
  reward_mean = flat_rewards.mean()
  reward_std = flat_rewards.std()
  return [(discounted_rewards - reward_mean) / reward_std
          for discounted_rewards in all_discounted_rewards]
```

To normalize all discounted rewards across all episodes, compute the mean and standard deviation of all the discounted rewards. Subtract the mean from each discounted reward and divide by the standard deviation. Let's try this function with two episodes:

```
discount_and_normalize_rewards(
    [[10, 0, -50], [10, 20]], discount_rate=0.8)
```

All actions from the first episode (first array) are considered bad because normalized advantages are all negative. This makes sense because the sum of the rewards is -40 (10 + 0 + -50). Conversely, the second episode actions (second array) are good because normalized advantages are positive. The sum of the rewards is 30 (10 + 20).

Train the Learner

As we know, RL uses a learning model and an environment space to enable an agent to learn how to solve a problem. So the agent learns how to solve a problem during the training process. So we can think of the process as training the learner.

Begin by defining a set of hyperparameters:

```
n_iterations = 150
n_episodes_per_update = 10
n_max_steps = 200
discount_rate = 0.95
```

We set 150 training iterations. Of course, you can tweak this number and the other values. Play ten episodes of the game per iteration, make each episode last at most 200 steps, and use a discount rate of 0.95.

Define an optimizer and loss function for the policy network:

```
optimizer = tf.keras.optimizers.Adam(learning_rate=0.01)
loss_fn = tf.keras.losses.binary_crossentropy
```

Use binary cross-entropy because we are training a binary classifier (two possible actions: left and right).

Generate a seed, clear previous models, and create a simple policy network:

```
tf.keras.backend.clear_session()
np.random.seed(0)
tf.random.set_seed(0)

model = Sequential([
  Dense(5, activation='elu', input_shape=[4]),
  Dense(1, activation='sigmoid'),
])
```

A policy network in reinforcement learning is very simple. Creating an environment space is the difficult part.

Train the learner as shown in Listing 14-9.

Listing 14-9. Train the Learner

```
env = gym.make('CartPole-v1')
env.seed(42);

for iteration in range(n_iterations):
  all_rewards, all_grads = play_multiple_episodes(
      env, n_episodes_per_update, n_max_steps,
      model, loss_fn)
```

```
  total_rewards = sum(map(sum, all_rewards))
  print('\rIteration: {}, mean rewards: {:.1f}'.format(
      iteration, total_rewards / n_episodes_per_update),
      end='')
  all_final_rewards = discount_and_normalize_rewards(
      all_rewards, discount_rate)
  all_mean_grads = []
  for var_index in range(len(model.trainable_variables)):
    mean_grads = tf.reduce_mean(
        [final_reward * all_grads[episode_index][step][var_index]
         for episode_index, final_rewards in enumerate(
             all_final_rewards)
           for step, final_reward in enumerate(
               final_rewards)], axis=0)
    all_mean_grads.append(mean_grads)
  optimizer.apply_gradients(
      zip(all_mean_grads, model.trainable_variables))

env.close()
```

To train the learner, begin by calling *play_multiple_episodes* (at each training iteration) to play the game ten times and return all the rewards and gradients for every episode and step. Next, call *discount_and_normalize_rewards* to compute each action's normalized advantage, which gives us a measure of how good or bad each action actually was in hindsight. For each trainable variable, compute the weighted mean of the gradients over all episodes and all steps weighted by the final_reward. The *final_reward* is each action's normalized advantage. End by applying the mean gradients using the optimizer, which tweaks the model's trainable variables to hopefully make the policy a bit better.

Note Training the learner takes some time. So be patient.

Render the Frames from the Reinforce Policy Gradient Algorithm

Render the frames:

```
frames_ra = render_policy_net(model)
```

Animate the Policy

Animate:

```
anim = plot_animation(frames_ra, repeat=True, interval=100)
anim
```

The pole appears to be a bit less wobbly. It's kind of amazing that the agent learned a better policy on its own!

Summary

We introduce the concept of RL and demonstrate it with a simple experiment. The code in the experiment is quite complex even for the simplest of environment spaces. The evolution of RL applied to real-world problems lies in the ability of designers to create environment spaces that realistically represent the world we live in.

Index

A

Adaptive instance normalization (AdaIN) layer, 297
Application programming interface (API), 2
Arbitrary Neural Artistic Stylization (ANAS)
  batch preparation, 302
  Google Drive, 299–300
  hub module, 303
  image stylization
    get images, 306
    pastiche creation, 309–310
    preprocessing function, 307
    processing images, 308
    reference dictionaries, 309
    style reference images, 307
  pastiche, 303–304
  preprocess images, 301
  pre-trained model, 298
  requisite libraries, 299
  RGB color model, 304
  style/transformer network, 296–297, 302
  TensorFlow lite module, 297
  visualization function, 305
Autoencoders
  break down, 208
  compile/train model, 205
  denoising process, 213–215
  dimensionality reduction, 208–209
  dropout, 215–217
  encoder model, 203–204
  GPU hardware accelerator, 202
  NumPy arrays, 203
  performance visualization function, 206
  reconstructions, 207
  representations/coding, 201
  scale data, 203
  stacked encoders, 203
  TensorFlow library, 202
  tying weights, 210–213
Automatic differentiation (AD), 281

B

Beans experiment
  Inception-v3
    compile/train, 182
    model creation, 180
    pre-trained layers, 182
    unfreezen layers, 182
    unseen test dataset, 183
  load model, 174
  metadata, 174
  resize/process images, 175
  transformation, 175
  visualization, 175
  Xception (*see* Xception model)

D. Paper, *State-of-the-Art Deep Learning Models in TensorFlow*, https://doi.org/10.1007/978-1-4842-7341-8

Beans loading
- checking shapes, 87
- compile/train model, 90
- convolutional layers, 89
- input shapes, 88
- I/O performance, 87
- meaning, 84
- metadata, 85
- multilayered CNN, 88
- pooling/stride, 89
- predictions, 90–91
- resize images, 87
- tf.data.Dataset.cache transformation, 88
- visualization, 85–86

C

CelebFaces Attributes Dataset (CelebA), 270
Central processing unit (CPUs), 128
Consumable input pipeline, 13–15
Convolutional autoencoders
- compile/train model, 225
- data loading, 220
- decoder, 224
- encoder creation, 224
- get training data, 222
- image displays, 222
- inspect shapes, 223
- libraries, 224
- metadata, 221
- preprocess images, 223
- reconstruction, 226
- visualization function, 225–226

D

Data augmentation
- function, 119–121
- image processing
  - augmentation function, 53
  - compile/train models, 110
  - display augmentations, 110
  - evaluation, 111
  - input pipeline, 109
  - model creation, 110
  - partial functions, 109
  - plot transformations, 54
  - predict() method, 112
  - predictions, 111–114
  - *random_crop* function, 108
  - training performance, 111
  - transformation, 48, 108
  - validation, 111
  - visualization, 113
- Keras API visualization, 44
- preprocessing layers
  - building models, 106
  - compile/train model, 107
  - evaluation, 107
  - transformation, 105
  - visualization, 108

Data augmentation
- definition, 37
- GPU hardware
  - accelerator, 38–39
- ImageGenerator class
  - compile/train model, 59
  - directories, 57
  - inspect data, 61
  - model creation, 58
  - multilayer CNN, 58
  - process flowers data, 56
  - rotate/flip images, 59
  - tf.data pipeline, 56
  - training data, 59–60
  - visualization, 62–63

image processing
bounding box, 49
brightness, 52
compile/train model, 55
contrast, 51
crop image, 49
display different images, 47
flip image, 50
function creation, 48
gamma encoding, 50
grabbing, 47
hue, 52
index variable, 47
model creation, 54
random files, 50
rotate image, 50
saturation, 52
train/test data, 54
transformation, 53
visualization, 47
Keras API, 39
compile/train model, 45
data information, 39
get number, 41
image shape/label, 41
input image, 42
model creation, 44
preprocessing layers, 42–43
scaling function, 42
split data, 40
train/test datasets, 40–41
visualization, 46
overfitting, 38
TensorFlow library, 38
Deep Convolutional generative adversarial network (DCGAN)
compile, 255
creation, 254
discriminator, 254
GANs modeling, 243
generator/discriminator, 252–253
image generation, 256
input pipeline, 255
large color images
discriminator model, 260–261
generator, 259
input pipeline, 258
massage data, 257
metadata, 256
NumPy arrays, 257
plotting function, 262
rescale function, 261
training loop function, 262
transform images, 258
visualization, 258
libraries, 252
reshape model, 255
tanh function, 253
train model, 255
Deep learning, 93
cats *vs.* dogs dataset, 93
compile/train model, 103
convolutional/pooling layers, 102
data augmentation
images, 108–115
preprocessing layers, 105–108
evaluate() method, 104
feature map, 103
generalization, 104
GPU hardware accelerator, 94
imshow() function, 100
input pipeline, 99
inspect, 98
inspect/visualization, 100
library, 94
metadata, 95-96

Deep learning (*cont.*)
NumPy arrays, 98
object loading, 95
resize/scale images, 99
rock_paper_scissors
compile/train model, 123
data augmentation creation, 119–121
input pipeline, 121
inspect data, 116
model creation, 122
preprocess, 117
TensorBoard configuration, 115, 125
Train/test sets, 116
training data, 118
visualization function, 118, 124
zoom augmentation, 120
split information, 96
shape model, 101
transform data, 99
visualization, 97, 104
Denoising model
compile/train model, 214
Gaussian noise, 214
objects (*see* Object detection)
meaning, 213
reconstructions, 215
training performance, 215
Dropout experiment
compile/train model, 217
configuration, 215
decoder, 216
reconstructions, 217
training process noisy, 215
visualization, 217

E

Ensorflow Probability (TFP) layers, 219

F

Flowers experiment
compile/train model, 195
generalization, 196
input pipeline, 193
one-hot encoding, 192–193
parse function, 192
pipeline/training parameters, 191
pre-trained models, 194–195
splits creation, 191–192
tf.data.Dataset, 193
TFRecord files, 190–191
training performance, 196
train/test datasets, 193–194

G, H

Generative adversarial networks (GANs)
data training, 248
DCGAN (*see* Deep Convolutional generative adversarial network (DCGAN))
discriminator, 245–248
Gaussian blurring, 250
generate images, 252
generator/discriminator model, 243
GPU hardware accelerator, 244
input shape/pipeline, 246, 248
iteration training, 250
libraries, 246
loss functions, 248
NumPy arrays, 245
scale model, 245
TensorFlow library, 244
training loop function, 249

train model, 251
untrained generator, 251
Google Cloud Storage (GCS)
convert text labels, 31
Fashion-MNIST dataset, 73
JPEG files, 29
TFRecord files
data model, 35–36
function creation, 33–35
process flowers, 31
read files, 31
set parameters, 32–33
train/test creation, 35
Graphics processing unit (GPUs), 128

I, J, K

Image generation
animation creation, 274
CelebA, 270–271
display interpolated image vectors, 274
function creation, 273
generate/display image, 273
interpolate hypersphere, 271–272
interpolation, 272
pre-trained model, 272
random vector, 276
single vector, 276
target latent vector, 277
Inception-v3
compile/train, 166
input pipeline, 165
iteration operates, 164
model creation, 166
plot function, 168
predictions, 168
reformat images, 165
training performance, 167
Input pipelines
automated workflow, 3
data cleaning, 2
from_tensor_slices() method, 4
machine learning, 1
manual workflow, 2
mechanics, 3

L

Latent vector/image arrays
float vector, 292
image vector, 293
latent space, 292
learning, 265–266
NumPy, 292
output tensor, 293
Loop learning experiment
animation creation, 283
AD, 281
closest latent function, 280
feature vector creation, 278–279
Google drive image, 287–289
image display, 279
latent vector, 278
learned images, 283
loss function algorithm, 281–284
model sessions, 282
regularization, 281
target image, 278
target image creation, 279
training function, 282
training loss, 282
uploaded image vector, 285–286
Wikimedia commons, 290–291

M

Machine learning (ML) models, 1, 341
Mean absolute error (MAE) reduction, 281
Mean squared error (MSE), 284
Memory dataset
  class labels definition, 12
  compile/train model, 17
  consumable input pipeline, 13–15
  dropout, 16
  I/O performance, 14
  libraries, 15–16
  load/inspect data, 8–9
  numerical label, 12
  plot creation, 13
  prefetching, 14
  scale data, 9–10
  scaling verification, 11
  shape model, 11
  shuffling data serves, 14
  tensor shapes, 11
  tf.data.Dataset objects, 9
  train/test tensors, 11
  visualization, 12
MobileNet-v2
  classification head, 159–160
  compile/train model, 160
  extraction, 159
  freeze/pre-trained layers, 159
  feature vector, 158
  ImageNet project, 155
  images/shapes displays, 157
  input pipeline, 158
  load flowers, 156
  metadata, 156–157
  plot predictions, 164
  predictions
    first batch images, 163
    inspect, 162
    np.argmax() function, 162
    test data, 162
  pre-trained model, 156
  reformat images, 158
  visual accuracy/loss performance, 161

N

Neural (fast) style transfer (NST)
  ANAS (*see* Arbitrary Neural Artistic Stylization (ANAS))
  GPU hardware accelerator, 298
  intermediate layers, 296
  pastiche, 295
  style transfer, 296
  TensorFlow library, 298
  TensorFlow Lite, 311–318

O

Object detection, 321
  bounding box, 323, 328–330
  BytesIO module, 327
  color channel, 322
  complex scenes
    download/preprocess function, 335
    paths, 337
    run option, 336
    source images, 338
    steps, 335
    URL protocol, 336
    Wikipedia Commons image, 338–339
  container function, 330
  detection *vs.* classification, 322–323
  digital image, 321
  function creation, 328

Google Drive, 332
GPU hardware
accelerator, 325
ImageDraw object, 329
natural scenes, 321–322
open images, 326
pre-trained model, 331
requisite libraries, 327
regression, 324–325
run detection, 333–334
temporary path, 333
TensorFlow library, 325
On-Device Portals (ODPs), 311

P, Q

Progressive Growing Adversarial
Networks (GAN)
environment experiments, 267
animation creation, 267
image processing libraries, 268
logging module, 268
function creation, 269
GPU hardware accelerator, 267
images (*see* Image generation)
imshow() function, 269
key innovation, 265
latent space
dimensions, 269–270
generative model, 266
learning, 265–266
observed pixel space, 266
latent vector/image arrays, 292–294
logging errors, 270
loop learning experiment, 278–291
TensorFlow library, 266
Python Imaging Library (PIL)
image, 299

R

Reinforcement learning (RL)
action space, 348–350
animation, 354
Cart-Pole game, 345
catastrophic forgetting, 343
computational approach, 341
cumulative reward, 341
environment, 346–347
episode, 350
GPU hardware accelerator, 344
libraries, 346
local/global optimum, 343
neural network, 343
neural network reward
policy, 351–352
NumPy array, 347
observation space, 351
OpenAI Gym, 345
optimization, 356
plot_animation function, 354
policy gradients algorithm
animation, 364
discount factor, 358
discount/normalize rewards, 360
frames, 364
function creation, 358
normalizes discounted rewards, 361
play multiple episodes
function, 359
optimization, 357
reinforcement learning, 362
training learning model, 362
predictions, 352–354
reinforcement learning, 342
render method, 347
rendered environment, 348
reward policy, 342, 355–357

Reinforcement learning (RL) (*cont.*)
sequential neural network, 352
TensorFlow library, 344
Rock-paper-scissors experiment
base model creation, 198
compile/train model, 198
dataset, 196
metadata, 197
process images, 197
test/train set data, 196
visualization, 197–199

S

Single shot detector (SSD), 324
Stanford Dogs experiment
base model creation, 187
compile/train model, 188
generalization, 190
image shape check, 186
input pipeline, 186
metadata, 184
MobileNets, 183
predictions, 189
preprocessing functions, 186
train set data, 184
Unfreezen layers, 190
visualization, 185
Supervised learning, 201

T

TensorBoard configuration, 115
TensorFlow Datasets (TFDSs)
auto-cache datasets, 80
beans loading (*see* Beans loading)
benchmark datasets, 80
compile/train, 21, 83
consumption (tf.data.Dataset objects), 20, 82
dataset builder, 65
dictionary, 68
fashion-MNIST
features dictionary, 73
load dataset, 72
metadata, 72
split information, 74
visualization, 74–75
GCS (*see* Google Cloud Storage (GCS))
GPU hardware accelerator, 66
input pipeline, 82
inspect shapes/pixel intensity, 19
Keras utility files, 21
library, 66
load option, 67–68
metadata, 68
models, 20–21
NumPy arrays, 18, 69–70
reloading object, 81
scale, 19
shape model, 82
single tensor, 81
slicing instructions
API data splits, 75
cross-validation, 77
ReadInstruction API, 78–80
strings, 76–78
softmax activation function, 21
tf.keras.preprocessing utility
compile/train model, 29
configuration, 27
CNN model, 27
get class names, 26
inspect tensors, 25
plot flower images, 26
scale images, 27

set parameters, 24
train/test sets, 25–27
tf.keras.utils.get_file utility, 22
tfds.core.DatasetBuilder class, 66
tuples, 69
visualization, 22–24, 71–72
TensorFlow library
data augmentation, 38
dataset
creation, 6, 7
numpy() method, 7
structure, 8
take() method, 7
Google Developers Codelabs, 4
GPU hardware accelerator, 5, 6
high-performance pipelines, 4
import, 5
input (*see* Input pipelines)
memory dataset (*see* Memory dataset)
notebook (Colab), 4–5
TensorFlow Lite
blending images, 316–317
crop images, 312
display function, 313
ODP platform, 311
pastiche, 316
pasticle PIL image, 318
pre-trained model, 311
style image, 314–316
TensorFlow Probability (TFP)
compile/train model, 239
decoder, 238
efficacy test, 240
ELBO equation, 240
Gaussian distribution, 236
Fashion-MNIST data, 235
input pipeline, 235
meaning, 234
model creation, 239
plotting function, 240
preprocess, 235
prior, 236
probabilistic models, 234
training batch, 236
VAE encoder, 237
visualization, 241
Tensor Processing Unit (TPU)
advantages, 128
cloud workloads
central processing unit, 128
double-precision arithmetic, 129
graphics processing unit, 128
computation, 132
detection, 130
digits experiment
consumption, 135
distribution strategy scope, 136
input pipeline, 135
libraries, 133
model data creation, 136
NumPy/SciPy, 134
preprocess data, 134
distribution strategy, 131
eager execution, 132
experiments, 133
fashion-MNIST experiment, 140
batch normalization, 142
inferences (predictions), 143–145
libraries, 141
NumPy tensors, 141
trained model, 143
transform datasets, 141
flowers dataset format, 145
data model, 150
display predictions, 152
function creation, 148

Tensor Processing Unit (TPU) (*cont.*)
inferences, 151
input pipeline, 149
set parameters, 147
TFRecord file, 146–148
train/test sets, 150
tf.data.Dataset, 148
hardware accelerator, 127
manual device placement, 131
MNIST experiment
create/compile model, 139
input pipeline, 138
repeat() method, 138
TFDS object, 137
TPUStrategy scope, 140
variable creation, 138
notebook configuration, 130
overview, 127
runtime type, 129
TensorFlow library, 129
Transfer learning
beans experiment, 173
image sizes, 174
inception, 180–183
input pipeline, 175
load model, 174
resize/process images, 175
visualization, 175
Xception model, 175
feature, 171
flowers, 190–196
GPU hardware accelerator, 154, 173
image classification models, 153
inception-v3, 164–169
library, 154
MobileNet-v2 (*see* MobileNet-v2)
pre-trained neural network, 153–154, 172
rock-paper-scissors, 196–199
stanford dogs, 183–190
TensorFlow library, 155, 172
Xception, 176–180
Tying weights experiment, 211–213

U

Unsupervised learning, 201

V, W

Variational autoencoder (VAE)
advantage, 219
decoder creation, 230
encoder creation, 229
GPU hardware accelerator, 220
interpolated coding, 234
latent/reconstruction loss, 231
plotting function, 232
random coding, 233
reconstructions, 232
sampling layer, 229
scale feature images, 228
semantic interpolation, 233
tensor shapes, 228
TensorFlow library, 220
Train/test images, 228
unsupervised technique, 227

X, Y, Z

Xception model
base model creation, 177
compile/train model, 178
pre-trained model, 176
unfrozen layers, 180
visualization function, 179